Springer Theses

Recognizing Outstanding Ph.D. Research

Aims and Scope

The series "Springer Theses" brings together a selection of the very best Ph.D. theses from around the world and across the physical sciences. Nominated and endorsed by two recognized specialists, each published volume has been selected for its scientific excellence and the high impact of its contents for the pertinent field of research. For greater accessibility to non-specialists, the published versions include an extended introduction, as well as a foreword by the student's supervisor explaining the special relevance of the work for the field. As a whole, the series will provide a valuable resource both for newcomers to the research fields described, and for other scientists seeking detailed background information on special questions. Finally, it provides an accredited documentation of the valuable contributions made by today's younger generation of scientists.

Theses are accepted into the series by invited nomination only and must fulfill all of the following criteria

- They must be written in good English.
- The topic should fall within the confines of Chemistry, Physics, Earth Sciences, Engineering and related interdisciplinary fields such as Materials, Nanoscience, Chemical Engineering, Complex Systems and Biophysics.
- The work reported in the thesis must represent a significant scientific advance.
- If the thesis includes previously published material, permission to reproduce this must be gained from the respective copyright holder.
- They must have been examined and passed during the 12 months prior to nomination.
- Each thesis should include a foreword by the supervisor outlining the significance of its content.
- The theses should have a clearly defined structure including an introduction accessible to scientists not expert in that particular field.

Indexed by zbMATH.

More information about this series at http://www.springer.com/series/8790

Marcel Rieger

Search for t̄tH Production in the H → bb̄ Decay Channel

Using Deep Learning Techniques with the CMS Experiment

Doctoral Thesis accepted by
RWTH Aachen University, Aachen, Germany

 Springer

Author
Dr. Marcel Rieger
CERN
Meyrin, Switzerland

III. Physics Institute A
RWTH Aachen University
Aachen, Germany

Supervisors
Prof. Martin Erdmann
III. Physics Institute A
RWTH Aachen University
Aachen, Germany

Prof. Alexander Schmidt
III. Physics Institute A
RWTH Aachen University
Aachen, Germany

ISSN 2190-5053   ISSN 2190-5061   (electronic)
Springer Theses
ISBN 978-3-030-65382-8   ISBN 978-3-030-65380-4   (eBook)
https://doi.org/10.1007/978-3-030-65380-4

This Springer imprint is published by the registered company Springer Nature Switzerland AG
The registered company address is: Gewerbestrasse 11, 6330 Cham, Switzerland

Supervisors' Foreword

The doctoral thesis of Dr. rer. nat. Marcel Rieger contains a number of ground-breaking innovations. In 2015, Marcel Rieger was one of the first to explore the technique of deep learning for particle physics. He developed deep neural networks for various applications and finally used a network for assigning particle collision events to physics categories in his data analysis. The network separates Higgs boson events, top quark events, events with both Higgs bosons and top quarks, and others. The separation power of deep learning surpasses previous analysis sensitivity. Another important innovation by Marcel Rieger concerns the efficient reproducibility of most demanding data analyses, as multiple investigations for analysis validity are mandatory. Marcel Rieger has developed a general workflow management system for user analyses. The user's control file reflects the design of the entire physical data analysis and contains all information necessary to automatically conduct and monitor the physical analysis at all levels. Using his technical innovations, Marcel Rieger analyzed collision events of the CMS experiment for the coupling between the Higgs boson and top quarks, focusing on the Higgs boson decay into two bottom quarks. Although the precision achieved in his analysis remains at the significance level of two standard deviations, Marcel Rieger's analysis provides the most important contribution to a combination with other CMS analyses that investigate other Higgs boson decays. Altogether, the proof of coupling of top quarks with Higgs bosons at a significance level of five standard deviations was conclusively demonstrated.

Overall, Marcel Rieger's dissertation documents these outstanding scientific achievements, for which he received the best possible grade for his doctorate at RWTH Aachen University. The CMS Experiment honored his achievements with the CMS Ph.D. Thesis Award 2019.

Aachen, Germany   Prof. Martin Erdmann
October 2020   Prof. Alexander Schmidt

Publications related to this Thesis

[1] M. Rieger, "Search for Higgs Boson Production in Association with Top Quarks and Decaying into Bottom Quarks using Deep Learning Techniques with the CMS Experiment", *Dissertation*, RWTH Aachen University (2019), http://dx.doi.org/10.18154/RWTH-2019-06415doi:10.18154/RWTH-2019-06415.

[2] CMS Collaboration, "Search for $t\bar{t}H \to b\bar{b}$ production in the decay channel with leptonic $t\bar{t}$ decays in proton-proton collisions at $\sqrt{s} = 13$ TeV", *JHEP* **03** (2019) 026, https://arxiv.org/abs/1804.03682arXiv:1804.03682 [hep-ex].

[3] CMS Collaboration, "Search for $t\bar{t}H$ production in the $H \to b\bar{b}$ decay channel with leptonic $t\bar{t}$ decays in proton-proton collisions at $\sqrt{s} = 13$ TeV with the CMS detector", *CMS Physics Analysis Summary CMS-PAS-HIG-17-026* (2018), http://cds.cern.ch/record/2308267cds:2308267.

[4] CMS Collaboration, "Observation of $t\bar{t}H$ production", *Phys. Rev. Lett.* **120** (2018) 231801, https://arxiv.org/abs/1804.02610arXiv:1804.02610 [hep-ex].

[5] CMS Collaboration, "Observation of Higgs boson decay to bottom quarks", *Phys. Rev. Lett.* **121** (2018) 121801, https://arxiv.org/abs/1808.08242arXiv:1808.08242 [hep-ex].

[6] M. Erdmann, B. Fischer and M. Rieger, "Jet-Parton assignment in $t\bar{t}H$ events using deep learning", *JINST* **12** (2017) P08020, https://arxiv.org/abs/1706.01117arXiv:1706.01117 [hep-ex].

[7] M. Erdmann, E. Geiser, Y. Rath and M. Rieger, "Lorentz Boost Networks: Autonomous Physics-Inspired Feature Engineering", *JINST* **14** (2019) P06006, https://arxiv.org/abs/1812.09722arXiv:1812.09722 [hep-ex].

[8] M. Erdmann, M. Rieger *et al.*, "Design and execution of make-like, distributed analyses based on Spotify's pipelining package Luigi", *J. Phys.: Conf. Ser.* **898** (2017) 072047, https://arxiv.org/abs/1706.00955arXiv:1706.00955 [physics.data-an].

[9] M. Erdmann, M. Rieger *et al.*, "Design Pattern for Analysis Automation on Distributed Resources using Luigi Analysis Workflows", *J. Phys.: Conf. Ser.* **1525** (2020) 012035.

[10] M. Erdmann, M. Rieger *et al.*, "A web based development environment for collaborative data analysis", *J. Phys.: Conf. Ser.* **523** (2014) 012021.

[11] [11] M. Erdmann, M. Rieger *et al.*, "Bringing experiment software to the web with VISPA", *J. Phys.: Conf. Ser.* **762** (2016) 012008.

[12] M. Erdmann, M. Rieger *et al.*, "Experiment software and projects on the web with VISPA", *J. Phys.: Conf. Ser.* **898** (2017) 072045, https://arxiv.org/abs/1706.00954arXiv:1706.00954 [physics.data-an].

Acknowledgments

A great number of people have contributed significantly to the realization of this work, for which I would like to take this opportunity to express my deepest gratitude.

First and foremost, my special thanks go to Prof. Dr. Martin Erdmann for the excellent supervision of this doctoral thesis. The support within and outside of the weekly working group meetings has always been a privilege to me and made the success of this work possible in the first place.

I would also like to warmly thank Prof. Dr. Alexander Schmidt for his collaboration and for assuming the role of the second referee of this dissertation.

For the productive cooperation, the exceptionally good atmosphere and the many fruitful discussions, I wish to thank all current and former members of the CMS working group lead by Prof. Dr. Martin Erdmann: Florian von Cube, Benjamin Fischer, Robert Fischer, Erik Geiser, Dennis Noll, Thorben Quast, Yannik Rath and David Schmidt. I would also like to thank all the members of my second working group VISPA, as well as those of the Auger working group. The numerous discussions about physics, deep learning and often far beyond were always a pleasure.

For proofreading this work, I would like to thank Prof. Dr. Martin Erdmann and Yannik Rath in particular.

Furthermore, I would like to express my deepest appreciation for the backing by the entire team of the III. Physics Institute. The technical support by the IT and Grid-Computing team headed by Dr. Thomas Kreß and Dr. Andreas Nowack contributed greatly to the success of this resource-intensive analysis. For their outstanding support at the institute, I wish to thank Dr. Markus Merschmeyer, Melanie Roder and Iris Rosewick.

Special thanks go to the Top and Higgs groups of the CMS experiment. I would like to thank the group and sub-group conveners for their support and helpful advice during this analysis: Dr. Maria Aldaya Martin, Dr. Florencia Canelli, Dr. Carmen Diez Pardos, Dr. Rainer Mankel, Dr. Paolo Meridiani, Dr. Roberto Salerno, Dr. Matthias Schröder and Dr. Caterina Vernieri. Moreover, I would like to thank all $t\bar{t}H(H \rightarrow b\bar{b})$ group members I closely collaborated with: Dr. Christian Contreras, Abhisek Datta, Karim El Morabit, Dr. Marco Harrendorf, Dr. Satoshi Hasegawa, Dr. Gregor Kasieczka, Philip Keicher, Dr. Wuming Luo, Dr. Joosep Pata, Dr. Aurelijus Rinkevicius, Andrej Saibel and Michael Wassmer.

Abschließend möchte ich mich bei meiner Familie und meinen Freunden für die immerwährende Unterstützung bedanken, die sie mir während meiner Doktorarbeit entgegen gebracht haben. Ohne sie wäre diese Arbeit nicht möglich gewesen.

Contents

1 Introduction 1
References 3

2 The $t\bar{t}H$ Process in the Standard Model of Particle Physics 5
2.1 The Standard Model of Particle Physics...................... 5
    2.1.1 Matter... 6
    2.1.2 Interactions 8
    2.1.3 Spontaneous Symmetry Breaking 13
2.2 The $t\bar{t}H$ Process at Hadron Colliders 17
    2.2.1 Hadron Collider Physics.......................... 17
    2.2.2 The Higgs Boson................................ 21
    2.2.3 The Top Quark 25
    2.2.4 Signal and Relevant Background Processes 28
    2.2.5 Results from Previous Searches 33
References 35

3 Experimental Setup... 41
3.1 The Large Hadron Collider 41
3.2 The Compact Muon Solenoid 45
    3.2.1 Tracking System 47
    3.2.2 Calorimeters 49
    3.2.3 Muon Detectors 51
    3.2.4 Trigger and Data Acquisition 53
    3.2.5 Luminosity Determination 54
3.3 Reconstruction of Physics Objects 55
    3.3.1 The Particle Flow Algorithm 55
    3.3.2 Muons 57
    3.3.3 Electrons...................................... 58
    3.3.4 Jets... 59

      3.3.5   Missing Transverse Energy   60
      3.3.6   *b*-Tagging   62
  3.4 Software and Libraries   63
      3.4.1   CMS Software Framework   64
      3.4.2   PXL and VISPA   65
      3.4.3   TensorFlow   67
      3.4.4   Luigi   68
  References   70

4 **Analysis Strategy**   73
  4.1 Physical Conditions   73
  4.2 Measurement Strategy   75
      4.2.1   Event Selection and Signal Extraction   76
      4.2.2   Event Categorization and Process Discrimination ..   78
  4.3 Technical Considerations   82
  References   83

5 **Analysis Technologies**   85
  5.1 Analysis Workflow Management   85
      5.1.1   Basic Concepts   86
      5.1.2   Remote Execution   89
      5.1.3   Remote Storage   89
      5.1.4   Environment Sandboxing   90
      5.1.5   Analysis Preservation   91
  5.2 Deep Learning and Neural Networks   92
      5.2.1   Mathematical Construction   92
      5.2.2   Neural Network Training   95
      5.2.3   Further Concepts   97
  5.3 Statistical Inference   99
      5.3.1   Likelihood Construction   100
      5.3.2   Inference Methods   103
      5.3.3   Saturated Goodness-of-Fit Test   107
  References   108

6 **Event Samples and Selection**   111
  6.1 Recorded Data Event Samples   111
  6.2 Simulated Event Samples   112
  6.3 Event Selection   117
      6.3.1   Vertex Selection and MET Filters   118
      6.3.2   Lepton Selection   119
      6.3.3   Trigger Selection   122
      6.3.4   Missing Transverse Energy Selection   123
      6.3.5   Jet and *b*-Tag Selection   123
      6.3.6   Selection Results   125
  6.4 Corrections to Simulated Events   126

6.5   Evaluation of Background Modeling   134
References   140

7   **Event Classification**   143
7.1   Input Variables and Validation of Correlations   144
7.2   Neural Network Architecture and Training   152
7.3   Binning Optimization   158
7.4   Evaluation of Training Results   159
References   170

8   **Measurement**   171
8.1   Systematic Uncertainties   171
8.1.1   Experimental Uncertainties   172
8.1.2   Theory and Background Modeling Uncertainties   180
8.1.3   Summary of Systematic Uncertainties   186
8.2   Measurement Results   186
8.2.1   Expected and Observed Limits   186
8.2.2 $t\bar{t}H$ Production Cross Section and Significance   188
8.2.3   Impact of Systematic Uncertainties   195
References   200

9   **Conclusion**   203
References   205

Appendix A: DNN Input Variables   207

Appendix B: DNN Output Distributions   213

Chapter 1
Introduction

The Standard Model of Particle Physics (SM) describes our current understanding of the constituents of matter and the forces that act between them. In 1964, a mechanism explaining the origin of masses of force-carrying particles was proposed by Robert Brout, François Englert, and Peter W. Higgs [1–3] as well as by Thomas W. B. Kibble, Carl R. Hagen, and Gerald Guralnik [4]. The mechanism postulates the existence of a new elementary particle and, by extension, also explains the masses of matter particles via Yukawa interaction. 48 years later, in 2012, the so-called Higgs boson was discovered in proton-proton collisions recorded by the ATLAS and CMS experiments at the Large Hadron Collider (LHC) [5, 6]. For their theoretical foundation, François Englert and Peter W. Higgs were jointly awarded the Nobel Prize in Physics in 2013.

The top quark is the heaviest elementary particle known to date with a mass comparable to that of a tungsten atom. Consequently, it involves a high coupling to the Higgs boson with a Yukawa coupling constant close to unity, which emphasizes its special role in the context of electroweak symmetry breaking. The accurate measurement of Higgs-top coupling is essential for probing predictions of the SM and to constrain various theories that reach beyond it. The most promising process to study is the production of a Higgs boson in association with a top quark pair ($t\bar{t}H$). Assuming a Higgs boson mass of $m_H = 125$ GeV and a top quark mass of $m_t = 172.5$ GeV, its cross section of $\sigma_{SM} = 507^{+35}_{-50}$ fb at a center-of-mass energy of $\sqrt{s} = 13$ TeV implies that a reasonable number of events was recorded in the proton-proton collisions during the past runs of the LHC [7].

In 2018, the CMS and ATLAS collaborations announced the observation of $t\bar{t}H$ production in agreement with the SM [8, 9]. Both collaborations combined analyses of various decay modes of the Higgs boson as well as multiple final states of the top quark pair system to measure an excess over expected background with a significance of more than five standard deviations. The CMS experiment measured the signal strength modifier $\mu = \sigma/\sigma_{SM}$, i.e., the cross section divided by its SM expectation, as $1.26^{+0.31}_{-0.26}$ using data at center-of-mass energies of $\sqrt{s} = 7, 8,$ and 13 TeV,

© The Author(s), under exclusive license to Springer Nature Switzerland AG 2021
M. Rieger, *Search for t̄tH Production in the H → bb̄ Decay Channel*, Springer Theses,
https://doi.org/10.1007/978-3-030-65380-4_1

corresponding to integrated luminosities of 5.1, 19.7, and 35.9 fb^{-1}, respectively [8]. The ATLAS experiment measured a cross section of 670^{+140}_{-130} fb using data taken at $\sqrt{s} = 13$ TeV, corresponding to an integrated luminosity of up to 79.8 fb^{-1} [9]. This corresponds to a signal strength modifier of $1.32^{+0.28}_{-0.27}$, which is in agreement with the measurement of the CMS experiment. Shortly afterwards, both collaborations also reported on the observation of Higgs boson decays into bottom quark pairs [10, 11]. After the observation of the couplings to vector bosons and tau leptons [12, 13], these results accomplished primary objectives of the Higgs physics program at the LHC as they provided the first decisive proofs of Higgs-quark coupling.

This thesis presents a search for $t\bar{t}H$ production in proton-proton collisions corresponding to an integrated luminosity of 35.9 fb^{-1} at a center-of-mass energy of 13 TeV recorded with the CMS detector in 2016. It focuses on Higgs boson decays into a pair of bottom quarks and leptonic top quark pair decays, i.e., one of the W bosons originating from the top quark decays into leptons while the other W boson decays either into a quark-antiquark pair (single-lepton) or into leptons (dilepton). Decays of W bosons that involve tau leptons are excluded. The analysis in the single-lepton channel, which dominates the overall sensitivity, contributed to the first observations of both $t\bar{t}H$ production [8] and Higgs boson decays into bottom quarks [10], performed with the CMS experiment. In combination with an alternative strategy in the dilepton channel, this analysis was published in Ref. [14].

The dominating background from top quark pair production, $t\bar{t}$, has a cross section of 832^{+46}_{-51} pb, which is $\sim$1600 times larger than the $t\bar{t}H$ cross section [15–27]. Especially events with additional heavy-flavor jets, such as $t\bar{t}+b\bar{b}$ and $t\bar{t}+c\bar{c}$, represent an important source of irreducible background. In addition, currently neither higher-order theoretical calculations nor experimental methods can constrain the expected rate of these contributions with an accuracy better than 35% [28–31].

A common approach for the separation of signal and background events is binary classification with multivariate analysis techniques (MVA). However, the existence of multiple different background processes indicates that a multi-class classification is favorable. This thesis introduces an event categorization scheme that is based on deep neural networks (DNN). Following this approach, events are either classified as signal or as one of the most significant background processes. Finally, the $t\bar{t}H$ cross section is extracted in a combined profile likelihood fit of the neural network output distributions to the recorded data.

The thesis is structured as follows. Section 2 briefly introduces the Standard Model of Particle Physics and characteristics of the $t\bar{t}H$ process such as the production at hadron colliders, decay, and experimental signature. Subsequently, Sect. 3 describes the experimental prerequisites of the analysis which includes the Large Hadron Collider, the CMS detector, algorithms applied to reconstruct physics objects, and software and libraries to perform the analysis. The analysis strategy, consisting of a summary of the physical conditions, the measurement methodology, and technical considerations, is presented in Sect. 4. Section 5 describes the technological key concepts that are employed to carry out the analysis at hand. The simulation of events as well as applied corrections are described in Sect. 6. It includes the selection of the analysis phase space and concludes with the evaluation of the overall background modeling. Subsequently, Sect. 7 explains the event classification approach that is based on deep learning techniques to separate signal from background contributions. The description of relevant systematic uncertainties and a study of their impact on the cross section measurement is given in Sect. 8, which is followed by final conclusions in Sect. 9.

Unit Convention

In the following sections, the convention of natural units is implicitly used, i.e., the physical constants.

- c, the speed of light in vacuum,
- $\hbar$, the reduced Planck constant, and
- ϵ_0, electric permittivity

are normalized to unity. Consequently, basic SI units are modified according to

- $[m] = \mathrm{GeV}$ (mass),
- $[t] = 1/\mathrm{GeV}$ (time), and
- $[s] = 1/\mathrm{GeV}$ (length).

References

1. Englert F, Brout R (1964) Broken symmetry and the mass of gauge vector mesons. Phys Rev Lett 13(9):321
2. Higgs PW (1964) Broken symmetries, massless particles and gauge fields. Phys Lett 12(2):132
3. Higgs PW (1964) Broken symmetries and the masses of gauge bosons. Phys Rev Lett 13(16):508
4. Guralnik GS, Hagen CR, Kibble TWB (1964) Global conservation laws and massless particles. Phys Rev Lett 13(20):585
5. ATLAS Collaboration (2012) Observation of a new particle in the search for the Standard Model Higgs boson with the ATLAS detector at the LHC. Phys Lett B 716:1. arXiv:1207.7214 [hep-ex]
6. CMS Collaboration (2012) Observation of a new boson at a mass of 125 GeV with the CMS experiment at the LHC. Phys Lett B 716:30. arXiv:1207.7235 [hep-ex]
7. LHC Higgs Cross Section Working Group (2017) Handbook of LHC higgs cross sections: 4. Deciphering the nature of the higgs sector. CERN Yellow Reports: Monographs (2017). arXiv:1610.07922 [hep-ph], cds:2227475
8. CMS Collaboration (2018) Observation of $t\bar{t}H$ production. Phys Rev Lett 120(23):231801. arXiv:1804.02610 [hep-ex]
9. ATLAS Collaboration (2018) Observation of Higgs boson production in association with a top quark pair at the LHC with the ATLAS detector. Phys Lett B 785:173. arXiv:1806.00425 [hep-ex]
10. CMS Collaboration (2018) Observation of Higgs boson decay to bottom quarks. Phys Rev Lett 121:121801. arXiv:1808.08242 [hep-ex]
11. ATLAS Collaboration (2018) Observation of $H \rightarrow b\bar{b}$ decays and VH production with the ATLAS detector. Phys Lett B 786:59. arXiv:1808.08238 [hep-ex]

12. CMS Collaboration (2017) Observation of the Higgs boson decay to a pair of τ leptons. Phys Lett B 779:283. arXiv:1708.00373 [hep-ex]
13. ATLAS Collaboration (2015) Evidence for the Higgs-boson Yukawa coupling to tau leptons with the ATLAS detector. JHEP 04:117. arXiv:1501.04943 [hep-ex]
14. CMS Collaboration (2019) Search for $t\bar{t}H$ production in the $H \rightarrow b\bar{b}$ decay channel with leptonic $t\bar{t}$ decays in proton-proton collisions at $\sqrt{s} = 13$ TeV. JHEP 03:026. arXiv:1804.03682 [hep-ex]
15. Beneke M et al (2012) Hadronic top-quark pair production with NNLL threshold resummation. Nucl Phys B 855:695. arXiv:1109.1536 [hep-ph]
16. Cacciari M et al (2012) Top-pair production at hadron colliders with next-to-next-to-leading logarithmic soft-gluon resummation. Phys Lett B 710:612. arXiv:1111.5869 [hep-ph]
17. Bärnreuther P, Czakon M, Mitov A (2012) Percent level precision physics at the tevatron: first genuine NNLO QCD corrections to $q\bar{q} \rightarrow t\bar{t}$. Phys Rev Lett 109:132001. arXiv:1204.5201 [hep-ph]
18. Czakon M, Mitov A (2012) NNLO corrections to top-pair production at hadron colliders: the all-fermionic scattering channels. JHEP 1212:054. arXiv:1207.0236 [hep-ph]
19. Czakon M, Mitov A (2013) NNLO corrections to top pair production at hadron colliders: the quark-gluon reaction. JHEP 1301:080. arXiv:1210.6832 [hep-ph]
20. Czakon M, Fiedler P, Mitov A (2013) The total top quark pair production cross-section at hadron colliders through $O(\alpha_s^4)$. Phys Rev Lett 110: 252004. arXiv:1303.6254
21. Czakon M, Mitov A (2014) Top++: a program for the calculation of the top-pair cross-section at hadron colliders. Comput Phys Commun 185:2930. arXiv:1112.5675 [hep-ph]
22. Botje M et al (2011) The PDF4LHC working group interim recommendations. arXiv:1101.0538 [hep-ph]
23. Martin A et al (2009) Parton distributions for the LHC. Eur Phys J C 63:189. arXiv:0901.0002 [hep-ph]
24. Martin A et al (2009) Uncertainties on α_s in global PDF analyses and implications for predicted hadronic cross sections. Eur Phys J C 64:653. arXiv:0905.3531 [hep-ph]
25. Lai H-L et al (2010) New parton distributions for collider physics. Phys Rev D 82:074024. arXiv:1007.2241 [hep-ph]
26. Ball RD et al (2013) Parton distributions with LHC data. Nucl Phys B 867:244. arXiv:1207.1303 [hep-ph]
27. Gao J et al (2014) CT10 next-to-next-to-leading order global analysis of QCD. Phys Rev D 89(3):03009. arXiv:1302.6246 [hep-ph]
28. CMS Collaboration (2018) Measurements of $t\bar{t}$ cross sections in association with b jets and inclusive jets and their ratio using dilepton final states in pp collisions at $\sqrt{s} = 13$ TeV. Phys Lett B 776:335. arXiv:1705.10141 [hep-ex]
29. Ježo T, Lindert JM, Moretti N, Pozzorini S (2018) New NLOPS predictions for $t\bar{t} + b$-jet production at the LHC. Eur Phys J C 78:502. arXiv:1802.00426 [hep-ph]
30. Garzelli MV, Kardos A, Trocsanyi Z (2015) Hadroproduction of $t\bar{t}b\bar{b}$ final states at LHC: predictions at NLO accuracy matched with Parton Shower. JHEP 03:083. arXiv:1408.0266 [hep-ph]
31. Bevilacqua G, Garzelli M, Kardos A (2017) $t\bar{t}b\bar{b}$ hadroproduction with massive bottom quarks with PowHel. arXiv:1709.06915 [hep-ph]

Chapter 2
The $t\bar{t}H$ Process in the Standard Model of Particle Physics

The endeavor of particle physics lies in the observation, formulation, and validation of rules that describe the properties of matter particles and the forces that act between them. The status of our current understanding is summarized in the Standard Model of Particle Physics (SM), which is discussed in the beginning of this section. It introduces the elementary matter particles consisting of quarks, leptons, and their antiparticles as well as the three interactions that act on subatomic scales. Gravity, as the fourth known fundamental force, is not described by the SM as its influence on fundamental particle processes is considered to be irrelevant at accessible energies. The "Higgs mechanism", which explains the origin of particle masses, is described subsequently. The second part of this section introduces the $t\bar{t}H$ event in the context of its production and decay characteristics at hadron colliders. The section continues with the specification of Higgs boson and top quark pair decay channels as studied in this thesis, and closes with a brief presentation of previous measurement results.

2.1 The Standard Model of Particle Physics

The SM is a consistent, finite theory based on quantum mechanics and special relativity. It succeeds in explaining most of the particle physics phenomena we observe with high accuracy. The nomenclature in its formalism, such as the concepts of "charge" and "current", is inspired by Quantum Electrodynamics (QED) for historical reasons. The SM describes electromagnetic, weak, and strong interactions between particles in terms of quantized fields following its nature of a quantum field theory. Mathematically, it is expressed by a Lagrangian density $\mathcal{L}$, which is required to be invariant under transformations of a governing symmetry group to preserve renormalizability [1]. Besides, it demands invariance under transformations of the Poincaré group, i.e., translational, rotational, and Lorentz-invariance in the Minkowski space, implying

© The Author(s), under exclusive license to Springer Nature Switzerland AG 2021
M. Rieger, *Search for t̄tH Production in the H → bb̄ Decay Channel*, Springer Theses,
https://doi.org/10.1007/978-3-030-65380-4_2

conservation of energy, momentum, and angular momentum according to Noether's theorem.

The formalism can be demonstrated using the QED Lagrangian with the $U(1)$ group,

$$\mathcal{L}_{\text{QED}} = \underbrace{\bar{\psi}(i\gamma^\mu D_\mu - m)\psi}_{\mathcal{L}_{int} \text{ (interaction term)}} - \underbrace{F_{\mu\nu}F^{\mu\nu}/4}_{\mathcal{L}_0 \text{ (field term)}} \quad (2.1)$$

$$\text{with} \quad D_\mu = \partial_\mu - ieA_\mu \quad (2.2)$$

$$\text{and} \quad F_{\mu\nu} = \partial_\mu A_\nu - \partial_\nu A_\mu, \quad (2.3)$$

where A_μ is the covariant field variable, m is the electron mass, e is the electric charge, γ^μ are the Dirac matrices, and ψ ($\bar{\psi} = \psi^\dagger \gamma^0$) is the spin $1/2$ field of the electron. Compared to the Lagrangian of a free (Dirac) particle, the covariant derivative (Eq. 2.2) is applied in the interaction term to achieve invariance under the local $U(1)$ gauge transformation

$$\psi \to \psi' = e^{iq\alpha(x)}\psi, \quad \text{given that} \quad A_\mu \to A'_\mu = A_\mu + \partial_\mu\alpha(x), \quad (2.4)$$

which introduces the gauge field A_μ of the photon. A mass term in the structure $m_\gamma A_\mu A^\mu$ would break the gauge symmetry and hence, the photon is massless. The quantity $\bar{\psi}\gamma^\mu\psi$ in the first summand of Eq. 2.1 is identified as the current j^μ. This leads to the definition of the "interaction" $e\, j^\mu A_\mu$, where e is taking the role of a coupling constant, usually transformed to the "coupling strength" $\alpha = e^2/4\pi$.

The field term $\mathcal{L}_0$ is quadratic in its field variables (A_μ) and can be solved exactly to obtain linear equations of motion. In combination with the interaction term $\mathcal{L}_{int}$, however, perturbation theory is required to obtain probability amplitudes for the involved physics processes. In doing so, the interaction term is considered a small perturbation of the free field. The effects of these perturbations can be approximated with expansions in leading (LO) or higher orders (NLO, NNLO, etc.), which can be employed to formulate transition probabilities using Fermi's golden rule. This calculation requires a description of an invariant scattering amplitude, which is often referred to as the "matrix element" and illustrated via Feynman diagrams.

2.1.1 Matter

Matter consists of elementary, point-like fermions with spin $1/2$. They are divided into quarks (u, d, c, s, t, b), leptons ($\nu_e, e, \nu_\mu, \mu, \nu_\tau, \tau$), and their corresponding antiparticles with opposite quantum numbers. An overview is presented in Table 2.1. Quarks with a positive electrical charge $Q = +2/3\, e$ are referred to as up-type quarks, (u, c, t), while the others with $Q = -1/3\, e$ are called down-type quarks (d, s, b). They are also subject to a color charge and participate in the strong, weak, and electromagnetic interactions. Charged leptons have an integer charge of $Q = -1\, e$ and interact weakly

Table 2.1 Elementary fermionic matter particles in the SM [2]. Each fermion has spin $1/2$ and can be associated to an antiparticle with opposite quantum numbers and, theoretically, identical mass. Quarks also exhibit one of three different color charges as described by QCD. The electron and muon masses are known to a relative accuracy of 10^{-8}. The electrical charge Q is given in units of the elementary charge e. The weak hypercharge can be obtained via $Y_W = 2 \cdot (Q - T_3)$

Type	Name	Symbol	Gen.	Charge	Weak Isospin	Mass
				$Q\,/\,e$	T_3	
Quark	Up	u	I	$+2/3$	$+1/2$	$2.2^{+0.5}_{-0.4}\,\text{MeV}$
	Down	d		$-1/3$	$-1/2$	$4.7^{+0.5}_{-0.3}\,\text{MeV}$
	Charm	c	II	$+2/3$	$+1/2$	$1.275^{+0.025}_{-0.035}\,\text{GeV}$
	Strange	s		$-1/3$	$-1/2$	$95^{+9}_{-3}\,\text{MeV}$
	Top	t	III	$+2/3$	$+1/2$	$173.0 \pm 0.4\,\text{GeV}$
	Bottom	b		$-1/3$	$-1/2$	$4.18^{+0.04}_{-0.03}\,\text{GeV}$
Lepton	electron neutrino	ν_e	I	0	$+1/2$	$< 2\,\text{eV}$
	Electron	e		-1	$-1/2$	$0.511\,\text{MeV}$
	Muon neutrino	ν_μ	II	0	$+1/2$	$< 0.19\,\text{MeV}$
	Muon	μ		-1	$-1/2$	$105.658\,\text{MeV}$
	Tau neutrino	ν_τ	III	0	$+1/2$	$< 18.2\,\text{MeV}$
	Tau	τ		-1	$-1/2$	$1776.86 \pm 0.12\,\text{MeV}$

and electromagnetically. In contrast, neutrinos are electrically neutral and thus, they are only subject to the weak interaction. The two groups can be further divided into so-called "generations", which are motivated by relations between their electrical charge, weak isospin component T_3, and mass hierarchy.

The masses of the elementary fermions range from a few eV for the electron neutrino up to the top quark mass of $m_t = 173.0\,\text{GeV}$, which is approximately as heavy as a tungsten atom. The structure of the mass hierarchy is not explained by the SM and entails the consequence that the matter we observe in nature mostly contains particles of the first generation due to the decay of heavier particles in higher generations. Neutrinos are not explicitly attributed a mass in the SM, but due to the observation of neutrino oscillations among different lepton flavors, they cannot be massless.

Quarks form color-neutral hadrons with two (mesons) or three (baryons) valence constituents, but bound states with four (tetraquarks) and five (pentaquarks) valence quarks were observed by the LHCb experiment at the LHC [3, 4]. With a lifetime of $\tau_t \approx 5 \cdot 10^{-25}\,\text{s}$, the top quark is the only quark that decays before hadronization occurs. Hence, it does not form any bound state and particle physicists are given the opportunity to study the decay of a bare quark. The electron is the only stable charged lepton. Muons decay after $\tau_\mu \approx 2.2 \cdot 10^{-6}$ s, which, however, can be considered as stable when produced in particle detectors at high energies. Tau lepton decays occur after $\tau_\tau \approx 2.9 \cdot 10^{-13}$ s and also involve hadrons since their mass of $m_\tau = 1776.86$ MeV is above the mass of the lightest charged meson, the pion $\pi^\pm$, with $m_{\pi^\pm} = 139.570$ MeV [2].

2.1.2 Interactions

The SM describes the effect of fundamental interactions as a quantum field theory with an exchange of force-mediating particles. The condition of local gauge invariance requires the introduction of gauge fields into the Lagrangian whose excitation modes are referred to as gauge bosons with an integer spin of 1. A summary of the twelve gauge bosons of the SM as well as the interactions they mediate and participate in is given in Table 2.2.

The Higgs boson, shown in the last row of Table 2.2, is not a mediator of any interaction. It is a scalar boson with spin 0 and emerges from the mechanism that explains how the $W^\pm$ and Z bosons acquire their masses within electroweak theory, while the photon remains massless. By extension, the mechanism also gives rise to the masses of quarks and charged leptons via additional Yukawa interactions, which predict the coupling strengths to be proportional to the fermion masses (cf. Sect. 2.1.3).

Gravitation is described by the general theory of relativity and applies on macroscopic scales. Due to its relative strength of 10^{-38} with respect to the strong interaction, its influence on fundamental particle processes is irrelevant and it is not considered in the SM.

As motivated in the introduction of this chapter, the SM is formulated as a Lagrangian density $\mathcal{L}_{\mathrm{SM}}$, governed by the symmetry product group

$$\mathcal{L}_{\mathrm{SM}} = \underbrace{SU(3)_C}_{QCD} \times \underbrace{SU(2)_L \times U(1)_Y}_{EW}. \tag{2.5}$$

Table 2.2 Force carrying gauge bosons and the Higgs boson in the SM [2]

Name	Symbol	Spin	Charge Q/e	Mass/GeV	Interaction	Subject to
Photon	γ	1	0	0	EM	EM
W boson (×2)	$W^\pm$	1	± 1	80.379 ± 0.012	Weak	Weak, EM
Z boson	Z	1	0	91.1876 ± 0.0021	Weak	Weak
Gluon (×8)	g	1	0	0	Strong	Strong
Higgs boson	H	0	0	125.18 ± 0.16	–	Self, weak

Here, the $SU(3)_C$ group describes the strong interaction, whereas $SU(2)_L \times U(1)_Y$ is associated to the electromagnetic and weak interactions. Their unification into the electroweak theory (EW) constitutes one of the most profound scientific achievements of the last century. The additional inclusion of QCD would lead to the so-called Grand Unified Theory (GUT). It implies that the relative strengths of the three interactions equalize at an energy scale of $\sim 10^{25}$ eV, which is experimentally unaccessible. The scale is inferred from searches for the predicted free proton decay, though it has not been observed so far. Thus, the formulation of a unified theory remains a major challenge in current physics research. In the subsequent paragraphs, brief explanations of the theoretical construction and interpretation of the SM are given, which loosely follow Refs. [2, 5–7] where it is described in greater detail.

Quantum Chromodynamics

The strong interaction is, as the name already implies, the strongest known interaction and causes quarks to form hadronic states and binds positively charged protons, despite their electromagnetic repulsion, to form the nuclei of atoms. The charge of the strong interaction is referred to as color charge. It exists in three different realizations, labeled "red", "green", and "blue", and is solely attributed to gluons and quarks, whereas antiquarks carry one of the corresponding anticolors.

The formulation of the strong interaction within quantum field theory is known as Quantum Chromodynamics (QCD), whose Lagrangian is constructed from the non-abelian, special unitary group color $SU(3)_C$, consisting of 3×3 matrices with determinant one. The eight generators, T_c with $c = 1 \ldots 8$, are expressed by the Gell-Mann matrices λ_c as $T_c = \lambda_c/2$ and do not commute,

$$[T_a, T_b] = i \sum_{c=1}^{8} f_{ab}^c T_c. \tag{2.6}$$

Here, f_{ab}^c are the complete antisymmetric $SU(3)_C$ structure constants. For a particular quark type with a color c, the Lagrangian reads

$$\mathcal{L}_{\text{QCD}} = \underbrace{\bar{\psi}_c(i\gamma^\mu D_\mu - m)\psi_c}_{\text{propagators}, \, gq\bar{q}\,\text{vertex}} - \underbrace{G_{\mu\nu}^c G_c^{\mu\nu}/4}_{g\,\text{self-coupling}} \tag{2.7}$$

$$\text{with} \quad D_\mu = \partial_\mu - ig_s T_c G_\mu^c \tag{2.8}$$

$$\text{and} \quad G_{\mu\nu}^c = \partial_\mu G_\nu^c - \partial_\nu G_\mu^c - g_s f_{ab}^c G_\mu^a G_\nu^b, \tag{2.9}$$

where g_s denotes the strong coupling constant and the covariant derivative D_μ introduces the eight massless gluon fields G_μ^c. It is invariant under the infinitesimal gauge transformations

$$\psi_c \to \psi_c' = e^{i\alpha_c(x)T_c}\psi_c \approx (1 + i\alpha_c(x)T_c)\psi_c \tag{2.10}$$

when the gluon fields transform as

$$G_\mu^c \to G_\mu^{c\prime} = G_\mu^c - \frac{1}{g_s}\partial_\mu\alpha_c(x) - f_{ab}^c\alpha_a(x)G_\mu^b. \tag{2.11}$$

By combining Eqs. 2.7–2.9, summarizing over all quark types and color states, and averaging over all possible final states, the structure of QCD interactions emerges. It entails a quark propagator, a gluon propagator, a $gq\bar{q}$ vertex with coupling constant g_s, as well as gluon self-couplings due to the non-vanishing commutators in Eq. 2.6, realized as triple and quadruple vertices with couplings of g_s and g_s^2, respectively. Inspired by QED, the coupling strength is interpreted as $\alpha_s = g_s^2/4\pi$.

Due to the non-abelian nature of the $SU(3)_C$ and the resulting gluon self-couplings, perturbation theory must be employed to derive physics predictions. Calculations of interactions with high momentum transfer Q^2 at short distances require regularization and renormalization, which methodically introduce an arbitrary mass scale μ_R. As inferred physics quantities must not depend on arbitrary chosen values, the scale dependence is compensated by introducing a "running coupling strength". A convenient parametrization of this behavior adds a scale parameter Λ_{QCD} (typically $\sim$200 MeV) that defines the momentum transfer at which the coupling is considered strong,

$$\alpha_s(Q^2) \propto \frac{1}{\ln Q^2/\Lambda_{\text{QCD}}^2}. \tag{2.12}$$

At energy scales equivalent to the mass of a Z boson the world average was measured as $\alpha_s(m_Z) = 0.1181 \pm 0.0011$ [2, 8, 9]. In particular, this dependence reveals that α_s decreases with increasing, high energy scales Q^2, a circumstance which is called "asymptotic freedom".

At low energy scales perturbation theory is not applicable and the distance behavior of the QCD potential becomes $\sim r$ with typical interaction lengths of $\mathcal{O}(10^{-15}\,\text{m})$. Therefore, free quarks cannot be observed in nature as they are either retained in bound, hadronic states or form new, color-neutral hadrons given a sufficient amount of energy to create pairs of quarks. They are subject to the so-called "color confinement". High energetic particle collisions, such as the proton-proton collisions generated at the LHC, induce quasi-free quarks in the range of the primary, hard physics process. Accounting for the propagation of the produced particles in space and the resulting decrease of the energy scale over time, the asymptotic freedom diminishes while color confinement gains relevance. This transition process is called "hadronization" and can be described by event generators only using phenomenological approaches.

Electroweak Theory

The electromagnetic interaction is the unification of electric and magnetic forces as described by Maxwell's equations. It is responsible for binding electrons and nuclei within atoms and molecules. The formulation as a relativistic field theory is known as Quantum Electrodynamics (QED), which is mediated by photons between particles carrying electric charge, i.e., quarks, charged leptons, and W bosons. Since the photon is massless, the range of electromagnetic interactions is infinite with a relative strength of $\sim 10^{-2}$ with respect to the strong interaction.

The weak interaction explains flavor-changing phenomena such as the β-decay of certain radioactive nuclei. In fact, it is the only known interaction that changes quark and lepton flavors, and that breaks particle-antiparticle symmetry (C), parity symmetry (P), and their combination (CP). The interaction is mediated by two electrically charged $W^{\pm}$ bosons, often referred to as *the* W boson, and the neutral Z boson, which were first observed by the UA1 and UA2 collaborations in 1983 [10, 11]. They couple to all elementary matter particles as well as to the Higgs boson. Their high masses of $m_W \approx 80.4\,\text{GeV}$ and $m_Z \approx 91.2\,\text{GeV}$ lead to a limited interaction range with typical interaction lengths of $\mathcal{O}(10^{-18}\,\text{m})$. Its relative strength compared to the strong interaction amounts to $\sim 10^{-13}$.

The electromagnetic and weak interactions are unified in the electroweak theory (EW), which marks a significant breakthrough in the development of the SM [12, 13]. Its Lagrangian is based on the product of two symmetry groups $SU(2)_L \times U(1)_Y$, which involve two different coupling constants g and g', as well as two sets of generators. The generator of $U(1)_Y$ is the weak hypercharge operator Y, whereas $SU(2)_L$ is generated by the three traceless, hermitian weak isospin operators T_a obeying the commutator rule

$$[T_a, T_b] = i \sum_{c=1}^{3} \epsilon_{abc} T_c, \qquad (2.13)$$

where ϵ_{abc} denotes the elements of the Levi-Civita tensor. The quantum numbers emerging for fermions are defined by eigenstates of these operators, which connect to the electric charge Q,

$$Q = T^3 + \frac{Y}{2}. \qquad (2.14)$$

Furthermore, the electroweak theory is a chiral theory, meaning that fermion fields are split into differently treated left- and right-handed components via projection operators

$$\psi \to \psi_L = \frac{1}{2}(1 - \underbrace{i\gamma_0\gamma_1\gamma_2\gamma_3}_{\gamma_5})\psi \qquad \bar{\psi} \to \bar{\psi}_L = \bar{\psi}\frac{1}{2}(1 + \gamma_5) \qquad (2.15)$$

$$\psi \to \psi_R = \frac{1}{2}(1 + \gamma_5)\psi \qquad \bar{\psi} \to \bar{\psi}_R = \bar{\psi}\frac{1}{2}(1 - \gamma_5). \qquad (2.16)$$

Components of left-handed fermions and right-handed antifermions have a chirality of -1 and constitute weak isospin doublets with (T, T_3) eigenvalues of $(1/2, \pm 1/2)$. Right-handed fermions and left-handed antifermions have a chirality of $+1$ and form weak isospin singlets with eigenvalue $T = 0$. As a result, electrically uncharged, right-handed neutrinos (left-handed antineutrinos) are not subject to any interaction and are thus not described in the SM.

The EW Lagrangian can be separated into three summands,

$$\mathcal{L}_{EW} = \mathcal{L}_{symm} + \mathcal{L}_{Higgs} + \mathcal{L}_{Yukawa}. \tag{2.17}$$

The symmetric part describes electroweak gauge and fermion fields, and reads

$$\mathcal{L}_{symm} = \underbrace{\sum_j i\bar{\psi}_L^j \gamma^\mu D_\mu^L \psi_L^j}_{T \text{ doublets}} + \underbrace{\sum_k i\bar{\psi}_R^k \gamma^\mu D_\mu^R \psi_R^k}_{T \text{ singlets (no neutrinos)}} - \frac{1}{4}\sum_{a=1}^3 W_{\mu\nu}^a W_a^{\mu\nu} - \frac{1}{4} B_{\mu\nu} B^{\mu\nu}$$

$$\tag{2.18}$$

with $W_{\mu\nu}^a = \partial_\mu B_\nu^a - \partial_\nu B_\mu^a - g\epsilon_{abc} W_\mu^b W_\nu^c$, $B_{\mu\nu} = \partial_\mu B_\nu - \partial_\nu B_\mu$,   (2.19)

and $D_\mu^L = \partial_\mu + ig\sum_{a=1}^3 T_L^a W_\mu^a + ig'\dfrac{Y_L}{2} B_\mu$, $D_\mu^R = \partial_\mu + ig'\dfrac{Y_R}{2} B_\mu$,   (2.20)

where W_μ^a and B_μ denote the gauge fields of the $SU(2)_L$ and $U(1)_Y$, respectively, and $D_\mu^{L,R}$ are the covariant derivatives for the left- and right-handed field components. Most notably, D_μ^R misses terms with $SU(2)_L$ gauge fields W_μ^a as the right-handed fermion fields are weak isospin singlets, i.e., $T_R^a \psi_R^k = 0$. The summations in Eq. 2.18 are applied for all left-handed weak isospin doublets j in the first, and for all right-handed singlets k in the second summand. Due to the definition of the covariant derivatives in Eq. 2.20, $\mathcal{L}_{symm}$ is invariant under the gauge transformations

$$\psi_L \rightarrow \psi_L' = e^{ig\alpha(x)T + ig'\beta(x)Y}\psi_L \tag{2.21}$$

$$\psi_R \rightarrow \psi_R' = e^{ig'\beta(x)Y}\psi_R. \tag{2.22}$$

The physical bosons (γ, Z, $W^\pm$) can be identified using mixtures of the gauge fields (W_μ^a, B_μ). When choosing the representation of the $SU(2)_L$ generators as $T_a = \sigma_a/2$, with the Pauli matrices σ_a, the eigenstates of the three gauge fields W_μ^a constitute the adjoint group representation in the form of unit vectors. It follows that

$$T_3 W_\mu^1 = i W_\mu^2, \qquad T_3 W_\mu^2 = -i W_\mu^1, \qquad T_3 W_\mu^3 = 0. \tag{2.23}$$

Since the $W^\pm$ bosons are observed to have an integer electrical charge of $Q = \pm 1\,e$ and are assigned a weak hypercharge of $Y = 0$, Eq. 2.14 leads to the requirement

$$T_3 W_\mu^\pm = \pm W_\mu^\pm, \tag{2.24}$$

which can be satisfied by defining

$$W_\mu^\pm = \frac{1}{\sqrt{2}}(W_\mu^1 \pm i W_\mu^2). \tag{2.25}$$

The γ and Z bosons can be motivated by comparing the electromagnetic interaction term in $\mathcal{L}_{\text{QED}}$ (Eq. 2.1),

$$e\, j_{\text{em}}^{\mu} A_{\mu} \overset{(2.14)}{=} e\left(\frac{1}{2} j_Y^{\mu} + j_3^{\mu}\right) A_{\mu} \tag{2.26}$$

with the corresponding interaction terms in $\mathcal{L}_{\text{symm}}$ (Eq. 2.18),

$$\frac{1}{2} g' j_Y^{\mu} B_{\mu} + g j_3^{\mu} W_{\mu}^3. \tag{2.27}$$

The interactions are equivalent when A_{μ}, and similarly Z_{μ} for weak neutral currents, are described by the orthogonal, normalized mixture of B_{μ} and W_{μ}^3,

$$\begin{pmatrix} A_{\mu} \\ Z_{\mu} \end{pmatrix} = \begin{pmatrix} \cos(\theta_W) & \sin(\theta_W) \\ -\sin(\theta_W) & \cos(\theta_W) \end{pmatrix} \cdot \begin{pmatrix} B_{\mu} \\ W_{\mu}^3 \end{pmatrix}. \tag{2.28}$$

The mixture represents a rotation by the weak mixing angle θ_W, also called "Weinberg" angle, given by $\tan(\theta_W) = g'/g$ and measured as $\sin^2(\theta_W) = 0.23122 \pm 0.00004$ in the SM [2].

The symmetric Lagrangian in Eq. 2.18 neither exhibits mass terms for fermion fields nor for gauge fields. While the photon is massless, W and Z bosons, as well as quarks and charged leptons are observed to be massive particles. In the SM, mass generation is described by the incorporation of a scalar field within $\mathcal{L}_{\text{Higgs}}$ (Eq. 2.17) that "spontaneously" breaks the electroweak symmetry. This mechanism is discussed in the following section.

2.1.3 Spontaneous Symmetry Breaking

In continuation of the introduction of the electroweak Lagrangian density in Eq. 2.17 of Sect. 2.1.2, the mechanism of electroweak symmetry breaking is discussed in the following. It explains how mass terms for W and Z bosons are generated by introducing a symmetric potential of a scalar field with an asymmetric ground state. The second part of this section describes the coupling of the scalar field to the previously introduced fermion fields, called Yukawa coupling, which gives rise to the mass terms of fermions.

The Higgs Lagrangian $\mathcal{L}_{\text{Higgs}}$, i.e., the second summand in Eq. 2.17, is defined as

$$\mathcal{L}_{\text{Higgs}} = (D_{\mu}\phi)^{\dagger}(D^{\mu}\phi) - V(\phi) \tag{2.29}$$

$$\text{with} \quad D_{\mu} = \partial_{\mu} + i\frac{g}{2}\sum_{a=1}^{3} \sigma_a W_{\mu}^a + i\frac{g'}{2} B_{\mu}, \tag{2.30}$$

following the generator representation chosen in Eq. 2.23, for a single weak isospin doublet ($Y = 1$, $T^3 = 1/2$) of two scalar fields,

$$\phi(x) = \begin{pmatrix} \phi^+(x) \\ \phi^0(x) \end{pmatrix}. \tag{2.31}$$

The peculiarity of the potential V in Eq. 2.29 is its symmetric shape that allows for asymmetric ground states,

$$V(\phi) = -\frac{\mu^2}{2}\phi^\dagger\phi + \frac{\lambda}{4}(\phi^\dagger\phi)^2, \tag{2.32}$$

which is illustrated in Fig. 2.1. In configurations where the constants μ^2 and λ are positive, the symmetry is spontaneously broken when the minimum of $V(\phi)$ is obtained for non-vanishing scalar fields ϕ, which is fulfilled at

$$\phi = \frac{1}{\sqrt{2}}\begin{pmatrix} 0 \\ \nu \end{pmatrix} \quad \text{with} \quad \nu = \frac{2\mu}{\sqrt{\lambda}}. \tag{2.33}$$

The vacuum expectation value (VEV) ν only depends on μ and λ. To perform perturbative calculations, an expansion of Eq. 2.31 around the VEV with $\phi^0 \to 1/\sqrt{2}(\nu + H + i\chi)$ is required. Without restricting generality, the unitary gauge $\mathrm{Re}(\phi^+) = \mathrm{Im}(\phi^+) = \chi = 0$ can be chosen owing to the symmetry of the ground state in $V(\phi)$. The scalar fields correspond to three so-called Goldstone   bosons,

Fig. 2.1 Illustration of the Higgs potential $V(\phi)$. While the potential itself is symmetric and satisfies gauge invariance, the ground state (purple sphere) is realized for all values on a circle with radius $\phi = \sqrt{2\mu^2/\lambda}$, effectively breaking the symmetry for non-vanishing fields ϕ

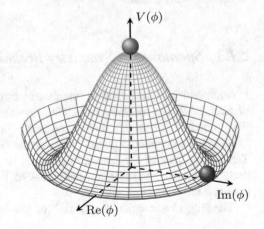

cannot be observed in nature, and transmute into the longitudinal components of the physical $W^\pm$ and Z bosons. This yields

$$\phi = \frac{1}{\sqrt{2}} \begin{pmatrix} 0 \\ \nu + H \end{pmatrix} \tag{2.34}$$

and the potential (Eq. 2.32), expanded in powers of the real Higgs field H, becomes

$$V = \mu^2 H^2 + \frac{\mu^2}{\nu} H^3 + \frac{\mu^2}{4\nu^2} H^4 = \frac{m_H^2}{2} H^2 + \frac{m_H^2}{2\nu} H^3 + \frac{m_H^2}{8\nu^2} H^4, \tag{2.35}$$

where the Higgs boson mass is identified as $m_H = \sqrt{2\mu^2}$. The potential also exhibits triple and quadruple Higgs boson vertices with couplings proportional to m_H^2/ν and m_H^2/ν^2, respectively.

Finally, the mass terms for the $W_\mu^\pm$ and Z_μ gauge fields emerge by applying the covariant derivatives to the first summand of the spontaneously broken Lagrangian $\mathcal{L}_{\text{Higgs}}$ (Eq. 2.29), when utilizing the mixing of (W_μ^a, B_μ) into $(W_\mu^\pm, Z_\mu, A_\mu)$ of Eqs. 2.25 and 2.28 that results in diagonal mass terms,

$$m_Z = \frac{1}{2}\sqrt{g^2 + g'^2}\,\nu, \quad m_W = \frac{1}{2}g\nu = \cos(\theta_W)m_Z. \tag{2.36}$$

Compatible with observations in nature, the photon field A_μ remains massless.

Fermion Masses

The Lagrangian densities discussed in the previous paragraphs do not contain mass terms for fermions. The symmetric term $\mathcal{L}_{\text{symm}}$ in Eq. 2.17 describes interactions of left- and right-handed fermion fields with electroweak currents, whereas the Higgs potential and the resulting mass terms for the $W^\pm$ and Z bosons are embedded in the second summand, $\mathcal{L}_{\text{Higgs}}$. The subsequent paragraphs address the last summand, $\mathcal{L}_{\text{Yukawa}}$, which introduces interactions between the scalar Higgs field and the spin-1/2 fermion fields to attribute masses to the latter.

The mass generation mechanism can be demonstrated with the left-handed weak isospin doublets and right-handed singlets of top and bottom quarks, $(\psi_L^t, \psi_L^b)^T, \psi_R^t$, and ψ_R^b, when neglecting flavor-mixing effects. Under this assumption, the Yukawa coupling terms become

$$\mathcal{L}_{\text{Yukawa},\,t/b} = -\lambda_t(\bar{\psi}_L^t, \bar{\psi}_L^b)\phi_u\psi_R^t - \lambda_b(\bar{\psi}_L^t, \bar{\psi}_L^b)\phi\psi_R^b \tag{2.37}$$

$$\text{with} \quad \phi_u = i\sigma_2\phi^*, \tag{2.38}$$

with the Yukawa coupling strengths $\lambda_{t,b}$ and the modified Higgs doublet ϕ_u for up-type quarks. Many extensions of the SM, such as the minimal supersymmetric SM (MSSM), do not contain this relation and hence, require two separate doublets. Applying the Higgs doublet of the spontaneously broken state (Eq. 2.34) yields

$$\mathcal{L}_{\text{Yukawa},\,t/b} = -\frac{1}{\sqrt{2}}(\lambda_t \bar{\psi}^t \psi^t + \lambda_b \bar{\psi}^b \psi^b)(\nu + H) \tag{2.39}$$

$$= -\underbrace{\lambda_t \frac{\nu}{\sqrt{2}}}_{m_t} \bar{\psi}^t \psi^t - \underbrace{\lambda_b \frac{\nu}{\sqrt{2}}}_{m_b} \bar{\psi}^b \psi^b - \underbrace{\lambda_t \frac{1}{\sqrt{2}}}_{m_t/\nu} \bar{\psi}^t \psi^t H - \underbrace{\lambda_b \frac{1}{\sqrt{2}}}_{m_b/\nu} \bar{\psi}^b \psi^b H.$$

$$\tag{2.40}$$

The emerging mass terms for top and bottom quarks m_t and m_b, respectively, are proportional to the Yukawa coupling constants $\lambda_{t,b}$. Since the top quark is measured to have by far the highest mass among all particles in the SM, precise experimental probes of Higgs-top couplings represent a powerful tool to test the mechanism of electroweak symmetry breaking and fermion mass generation.

The inclusion of all three quark generations requires a generalized formulation of the Yukawa coupling terms. In the following, this generalization is discussed in the scope of quark masses. The approach can be adapted to leptons in a similar manner. Equation 2.39 can be extended to

$$\mathcal{L}_{\text{Yukawa}} = -\bar{\psi}_L \mathcal{M} \psi_R - \bar{\psi}_R \mathcal{M}^\dagger \psi_L - \bar{\psi}_L \frac{\mathcal{M}}{\nu} \psi_R H - \bar{\psi}_R \frac{\mathcal{M}^\dagger}{\nu} \psi_L H, \tag{2.41}$$

where $\mathcal{M}$ is the non-diagonal fermion mass matrix as generated by the coupling to the Higgs field. The matrix can be diagonalized using two unitary transformations,

$$\mathcal{M}_{\text{diag}} = V_L^\dagger \, \mathcal{M} \, U_R, \tag{2.42}$$

which is equivalent to a basis change of up- and down-type quarks given by the left-handed doublets ψ_L and right-handed singlets ψ_R in the symmetric term of the Lagrangian (Eq. 2.18),

$$\psi_L^u \to \psi_L^{u\prime} = V_L^u \, \psi_L^u, \qquad\qquad \psi_L^d \to \psi_L^{d\prime} = V_L^d \, \psi_L^d \tag{2.43}$$

$$\psi_R^u \to \psi_R^{u\prime} = U_R^u \, \psi_R^u, \qquad\qquad \psi_R^d \to \psi_R^{d\prime} = U_R^d \, \psi_R^d. \tag{2.44}$$

This transformation neither changes the couplings to neutral gauge fields nor to the scalar Higgs field. It does, however, induce a mixing of mass eigenstates to interactions between left-handed fermion doublets and charged gauge fields. This mixing is represented by the Cabibbo–Kobayashi–Maskawa (CKM) matrix [2],

$$V_{\text{CKM}} = \left(V_L^u\right)^\dagger V_L^d = \begin{pmatrix} V_{ud} & V_{us} & V_{ub} \\ V_{cd} & V_{cs} & V_{cb} \\ V_{td} & V_{ts} & V_{tb} \end{pmatrix}, \tag{2.45}$$

and summarizes the projection of mass eigenstates into fermion fields at flavor-changing Wqq' vertices. By construction, V_{CKM} is unitary ($V^\dagger V = 1$), which constrains its free parameters to three mixing angles and a complex phase. The observation of CP violation in weak interactions implies that this phase is different from zero. Moreover, the unitarity can be exploited to achieve an over-constrained, global fit combining all separate measurements of the CKM matrix elements, which results in amplitudes $|V_{ij}|$ of [2]

$$
V_{\text{CKM}} = \begin{pmatrix} 0.97446 \pm 0.00010 & 0.22452 \pm 0.00044 & 0.00365 \pm 0.00012 \\ 0.22438 \pm 0.00044 & 0.97359^{+0.00010}_{-0.00011} & 0.04214 \pm 0.00076 \\ 0.00896^{+0.00024}_{-0.00023} & 0.04133 \pm 0.00074 & 0.999105 \pm 0.000032 \end{pmatrix}.
$$

(2.46)

2.2   The $t\bar{t}H$ Process at Hadron Colliders

After the description of the particle content of the SM and the theoretical construction of the interactions between them in Sect. 2.1, this section introduces the $t\bar{t}H$ process, which is subject of the analysis presented in this thesis. The section starts with a phenomenological description of proton-proton collision physics and explains the concepts employed to model collision events with characteristics that are comparable to real measurement conditions. The two subsequent sections discuss features of the Higgs boson and the top quark, respectively, such as production processes at hadron colliders, signatures of their decay modes, and latest results of experimentally measured properties. Following this, the $t\bar{t}H$ process and important background contributions from other physics processes are presented. The production of a top quark pair with additional heavy-flavor jets, e.g. from $t\bar{t}+b\bar{b}$ or $t\bar{t}+c\bar{c}$ processes, is particularly important as it constitutes the major background to this analysis. Additional jets can either emerge within or close to the hard process initiated by quark or gluon radiations, or they arise in later stages of the jet formation, namely parton showering and hadronization. The theoretical description of these processes is challenging and thus, distinct and model-independent process definitions based on measurable quantities are necessary for the study of event topologies. The section closes with the description of the $t\bar{t}H$ decay channels that are studied in this analysis, followed by a summary of experimental results measured in previous analyses performed by the CMS and ATLAS collaborations.

2.2.1   Hadron Collider Physics

Particle colliders are designed to accelerate two beams of particles in opposite directions up to a certain kinetic energy and induce collisions whose products can be studied in adjoining particle detector experiments. In contrast to linear variants, circular colliders have the advantage that particles pass accelerating sections once every revolution to gradually increase their energy. On the downside, particles emit synchrotron radiation with an energy loss proportional to $(E/m)^4$ when undergoing radial acceleration to maintain a circular trajectory. Since the energy loss of electrons poses a technically limiting factor at high center-of-mass energies, $\sqrt{s} = 2 \times E_{beam}$, colliders such as the LHC use protons, which results in a reduction by a factor of $(m_p/m_e)^4 \sim 10^{13}$.

Protons are not elementary particles but hadrons and consist of three valence quarks, specifically two up quarks and one down quark, and a sea of gluons and quark-antiquark pairs. The term "parton" is commonly used throughout literature to describe a generic hadron constituent. The collision of two protons must be understood as the interaction of two partons i and j that carry only a fraction of the total proton momenta x_1 and x_2, respectively. These fractions are described by parton distribution functions (PDFs) and exhibit a dependence on the momentum transfer Q^2 and on the particular flavor of the involved parton, i.e., u, d, c, s, b, or g. Exemplary PDFs for the NNPDF3.0 set with $\alpha_s = 0.118$ are shown in Fig. 2.2 for a low momentum transfer $Q^2 = 10\,\text{GeV}^2$ on the left-hand side, and for a high momentum transfer $Q^2 = 10^4\,\text{GeV}^2$ on the right-hand side [2, 14].

The figures show separate distributions for the valence quarks u_v and d_v, and for partons associated to the sea. In accordance with the general perception, there is a high probability for valence quarks to carry a large fraction of the proton momentum. Sea quarks predominantly contribute with low momentum fractions, whereas gluons also yield a significant amount in higher regimes of x. At high values of Q^2 (Fig. 2.2, right), the contribution of sea partons gains relevance. In particular, values of the

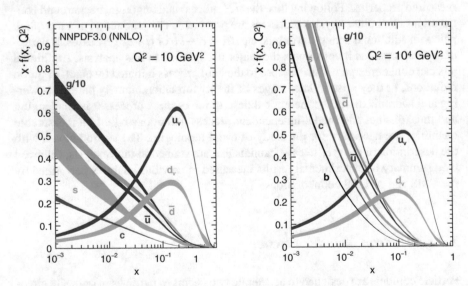

Fig. 2.2 Parton distribution functions $x \cdot f(x, Q^2)$ for the PDF set NNPDF3.0 with $\alpha_s = 0.118$ [2]. The momentum transfer scales Q^2 are chosen as $10\,\text{GeV}^2$ (left) and $10^4\,\text{GeV}^2$ (right). The higher scale also involves bottom quarks. Contributions of the valence quarks u_v and d_v are shown in addition to the total up and down quark functions, respectively. Gluon distributions are scaled by a factor of 0.1

gluon PDF exceed those of sea and valence quarks up to fractions x of approximately $\sim$10%, which, in further consequence, causes processes such as $t\bar{t}H$ production to be mainly induced by gluon-gluon interactions (cf. Sect. 2.2.4).

PDFs are typically inferred by measuring hadron structure functions in deep-inelastic lepton-hadron scattering processes. Among others, measurements were conducted in electron-proton collisions at the Hadron-Elektron-Ring-Anlage (HERA) accelerator [15, 16], or in fixed-target neutrino-nucleus scattering studies with the CCFR/NuTeV experiments [17]. Various groups performed global fits of the sum of weighted PDFs to hadron structure functions measured by different experiments [2]. This analysis uses the NNPDF3.0 set with $\alpha_s = 0.118$ as recommended by the PDF4LHC group for the second run of the LHC [18, 19].

The total cross section of a process $pp \rightarrow X$ can be formulated using the factorization theorem,

$$\sigma(pp \rightarrow X \mid \sqrt{s}) = \sum_{i,j}^{\text{partons}} \int \hat{\sigma}(ij \rightarrow X \mid \sqrt{\hat{s}}, \mu_R^2, \mu_F^2) \, f_i(x_1, \mu_F^2) \, f_j(x_2, \mu_F^2) dx_1 dx_2,$$

(2.47)

which sums over all possible parton combinations i, j, and separates PDFs from cross sections at parton-level $\hat{\sigma}$ at a reduced center-of-mass energy $\sqrt{\hat{s}} = \sqrt{x_1 x_2 s}$. The situation is visualized in Fig. 2.3. In addition, the factorization scale μ_F^2 is introduced to control the description of quark-gluon interactions. Interactions with $Q^2 < \mu_F^2$ are absorbed directly into the PDF, whereas those with high momentum transfer, $Q^2 > \mu_F^2$, are to be considered within the parton-level cross section. The renormalization scale μ_R^2 defines the scale for the strong coupling constant α_s calculation (cf. Sect. 2.1.2). Typically, a consistent scale $\mu_R = \mu_F$ is selected, which resembles the momentum scale of the hard scattering process, e.g. m_t for processes involving top quarks. The fixation to a singular value is accounted for in experimental applications with systematic uncertainties using independent variations of μ_R and μ_F by factors of $1/2$ and 2.

Fig. 2.3 Diagram of the factorization in a hard scattering process between two hadrons into the parton-level cross section $\hat{\sigma}$ and the parton distribution functions $f(x, \mu_F^2)$ describing the probability for partons to carry the hadron momentum fraction x at the factorization scale μ_F^2 [20]

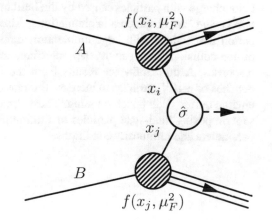

After applying the factorization theorem to describe hard scattering processes, the energy scale of emerging particles typically exceeds the QCD renormalization scale, $Q^2 > \mu_R^2$, so the strong coupling constant α_s is sufficiently small for perturbative calculations to converge. For quarks, the evolution of the scale down to a certain cutoff value is described by "parton showering". It introduces recursive splittings $q \rightarrow qg$ and subsequently $g \rightarrow q\bar{q}$ in a well-defined order of the involved partons based on their splitting probability, which can be calculated in any order of α_s. Particles with energies below the cutoff scale are not considered for splittings anymore. This approach is usually implemented in Monte Carlo (MC) event generators [21]. While the prescription of matrix elements obtained from perturbative calculations is most effective for hard and large-angle emissions, parton showering yields superior efficiency for soft and collinear radiations. Therefore, a "matching" algorithm is employed to mediate between the two descriptions based on kinematic variables such as emission angle and momentum [22, 23].

Finally, when the energy scale falls below the QCD renormalization scale, $Q^2 < \mu_R^2$, and strong coupling must be considered, perturbative calculations loose relevance. To describe this "hadronization" process, event generators must rely on phenomenological models, such as the Lund string model [24], which are subject to tuning. Thereafter, unstable hadrons are considered to decay until only stable, final-state particles remain. The result is a jet of collimated, stable hadrons as well as leptons from decays of unstable hadrons. The combined four-momentum of all jet constituents is supposed to resemble the four-momentum of the initial quark. Therefore, jet clustering algorithms represent a powerful tool to infer kinematic quantities of the hard scattering process. Typical methods are the k_t, anti-k_t, and Cambridge-Aachen (CA) clustering algorithms [25–28]. An illustration of the event modeling procedure is shown in Fig. 2.4.

Additional, residual effects need to be taken into account in the modeling of proton-proton collisions. Apart from the partons that contribute to the hard scattering process, remnants of the colliding protons must form color-neutral hadrons under influence of QCD color confinement (cf. Sect. 2.1.2). As they potentially exchange color charge with particles caused by the hard interaction, the effects of this "underlying event" are considered in simulations. Also, it is possible that multiple parton-parton interactions with large momentum transfer occur within the same proton-proton collision. They can overlap, interfere, and lead to indistinguishable detector responses. A third influence results from the fact that protons are accelerated in bunches of $\sim 10^{11}$ particles to increase the rate of interesting physics processes that emerge during each bunch crossing. These "pileup" collisions can be simulated by superimposing final-state particles of minimum-bias interactions that match beam parameters and luminosity conditions.

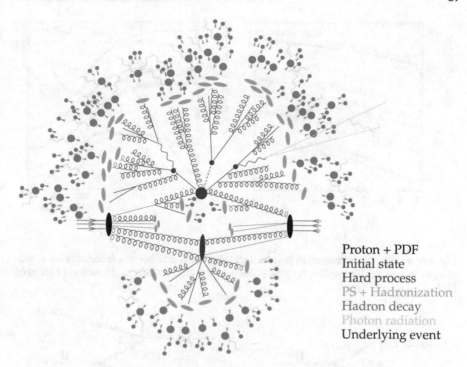

Fig. 2.4 Illustration of a $t\bar{t}H$ event as generated by a Monte Carlo event generator [29]. Quarks and gluons in the initial state (blue) contribute with fractions of the energy of the colliding protons according to PDFs. The hard interaction and additional hard QCD emissions are shown in red. An underlying event interaction is visible in purple. After parton showering and hadronization of intermediate hadrons (light green), the decay into the final state takes place (dark green). Additional photon radiations are added in yellow

2.2.2 The Higgs Boson

The Higgs boson represents the quanta of the scalar field that is introduced in the discussion of electroweak symmetry breaking in Sect. 2.1.3. The mechanism explains how W and Z bosons acquire their mass while the photon remains massless. By extension, fermion masses are generated via Yukawa interactions between the scalar Higgs field and the spin-1/2 fermion fields. In the SM, the Higgs boson is expected to be electrically neutral and scalar, i.e., it is free of spin and CP-even ($J^P = 0^+$), with a theoretical decay width of about 4 MeV [2].

Production Processes

At proton-proton colliders such as the LHC, four significant production processes exist at accessible center-of-mass energies, which are described in the following. The energy-dependent cross sections are presented in Fig. 2.5(a) for a Higgs boson mass of $m_H = 125$ GeV. Predictions are calculated in different orders of perturbation theory whereas their uncertainties stem from variations of PDF models, renormal-

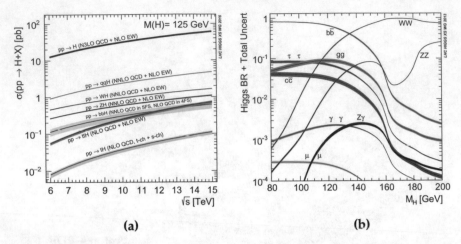

(a)   **(b)**

Fig. 2.5 a Higgs boson production processes and their cross sections as a function of the center-of-mass energy $\sqrt{s}$ [30]. **b** The branching ratios of the Higgs boson as a function of its mass [30]

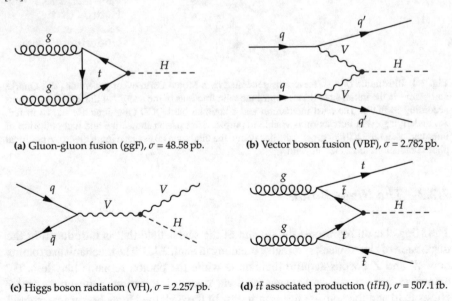

(a) Gluon-gluon fusion (ggF), $\sigma = 48.58\,\text{pb}$.   **(b)** Vector boson fusion (VBF), $\sigma = 2.782\,\text{pb}$.

(c) Higgs boson radiation (VH), $\sigma = 2.257\,\text{pb}$.   **(d)** $t\bar{t}$ associated production ($t\bar{t}H$), $\sigma = 507.1\,\text{fb}$.

Fig. 2.6 Feynman diagrams and cross sections of the four most significant Higgs boson production processes at hadron colliders at $\sqrt{s} = 13\,\text{TeV}$ and $m_H = 125\,\text{GeV}$ [30]

ization scales, and factorization scales (cf. Sect. 2.2.1). The corresponding Feynman diagrams and cross sections at $\sqrt{s} = 13\,\text{TeV}$ are depicted in Fig. 2.6.

With a cross section of $\sigma = 48.58\,\text{pb}$ at N3LO, the dominant production mechanism originates from the fusion of two gluons and a virtual loop of heavy quarks (ggH, Fig. 2.6a) [30]. Due to their high masses, the top quark and, to a lesser extent, the bottom quark mainly contribute to the loop calculation.

The second most common production process, namely vector boson fusion, exhibits a cross section that is one order of magnitude smaller with $\sigma = 2.782\,\text{pb}$ (VBF, Fig. 2.6b) [30]. The value is calculated at approximate NNLO with electroweak accuracy corrections at NLO (NNLO QCD + NLO EW). Here, two Z bosons or oppositely charged W bosons are radiated from quarks in the initial state and produce a Higgs boson. The accompanying, remnant quarks tend to be spatially separated and their topology can be exploited to identify this process.

In the third production process, a Higgs boson is radiated off a Z or W boson (VH, Fig. 2.6b) [30]. Compared to VBF, this so-called Higgs-Strahlung contains just one $Vq\bar{q}/q'$ vertex, which should in general lead to a higher cross section. However, the initial state with an antiquark from the parton sea is less probable as predicted by PDFs (cf. Sect. 2.2.1) and results in a comparable cross section of $\sigma = 2.257\,\text{pb}$ (NNLO QCD + NLO EW).

With a cross section of $\sigma = 507.1\,\text{fb}$ (NLO QCD + NLO EW), two orders of magnitudes below ggH, the least common of the four processes is the production of a Higgs boson in association with a top quark pair ($t\bar{t}H$, Fig. 2.6d) [30]. As opposed to ggH, the coupling involves non-virtual top quarks and allows for a direct measurement of the Higgs-top Yukawa interaction in the light of electroweak symmetry breaking. In this context, the production of a Higgs boson with a single top quark (bottom curve in Fig. 2.5a) is potentially a second candidate, but due to negative interference of leading-order diagrams, its cross section is drastically reduced. Therefore, the analysis presented in this thesis focuses on $t\bar{t}H$ production.

Decay Channels

The following paragraphs discuss the possible Higgs boson decay modes at a mass of $m_H = 125\,\text{GeV}$. They can be divided into vector boson, fermion, and loop-induced decays. A visualization of the branching ratios (BR) as a function of m_H is shown in Fig. 2.5b.

The decay into a pair of bottom quarks, $H \to b\bar{b}$, yields the highest branching ratio (58.2%). Among the other decays into fermions, this is followed by $H \to \tau^+\tau^-$ (6.72%). The detection of these channels is challenging due to the vast contamination of jet backgrounds and therefore, analyses additionally consider unique event signatures of VH or $t\bar{t}H$ production. The decay channel $H \to c\bar{c}$ (2.9%) is even more challenging and highly depends on sophisticated algorithms for reconstructing the initial quark flavor. Further decays into fermions ($u\bar{u}$, $d\bar{d}$, $s\bar{s}$, $\mu\bar{\mu}$) have vanishingly low branching ratios and are experimentally dominated by backgrounds.

Decays into vector bosons are restricted to pairs of W (21.4%) and Z bosons (2.6%). Despite their high masses, these decay channels are suppressed since the Higgs boson is too light to decay into two vector bosons on their mass shell, i.e., the relation $E^2 = p^2 + m^2$ is not fulfilled, which is often indicated by an asterisk. Further decays into purely leptonic final states, such as $H \to ZZ^* \to l^+l^-l^+l^-$, lead to very clear signatures with reasonably small background contributions and are well suited for measuring Higgs boson properties.

Similar to the Higgs boson production via fusion of massless gluons, decays into bosons can be induced by (triangle-)loops. Quark loops, mostly involving top quarks, effectively enable the decay channels $H \to gg,\ Z\gamma,\ \gamma\gamma$. The latter, $H \to \gamma\gamma$, is also realizable through a W boson loop and has a rather small branching ratio (0.2%). Owing to its experimental mass resolution of 1–2% [2], it still had a significant impact on the first observation of Higgs boson production in 2012 [31, 32]. Although the decay into gluons has a sizable branching ratio (8.2%), it is unlikely to be detected at hadron colliders.

Measured Properties

Since its first observation in 2012, various measurements of the Higgs boson's properties were conducted by the ATLAS and CMS collaborations. All of the aforementioned production processes could be observed with significances of at least five standard deviations above expected background contributions [31–35]. Furthermore, decays into pairs of vector bosons, WW^* [36, 37], ZZ^* [38], and $\gamma\gamma$ [39, 40], as well as into fermions, $\tau^+\tau^-$ [41] and $b\bar{b}$ [33, 42], were observed. Fig. 2.7 summarizes current results of Higgs boson coupling measurements as performed by the CMS collaboration. It shows central values and uncertainties on signal strength modifiers of particular production processes subdivided into accessible decay channels. Within their measurement uncertainties, all results are compatible with predictions

Fig. 2.7 Measurement of signal strength modifiers μ_i^f of Higgs boson production processes i subdivided into accessible decay channels f, performed by the CMS collaboration at $\sqrt{s} = 13\,\mathrm{TeV}$ [43]. The SM prediction of $\mu_i^f = 1$ is represented by the vertical dashed line. Black markers represent the best fit values, whereas the horizontal lines denote CL intervals corresponding to one standard deviation. Hatched areas indicate signal strengths which are restricted to non-negative values

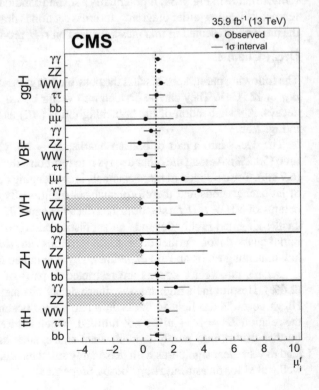

of the SM. Therefore, it could be clearly established that the observed Higgs boson is electrically neutral.

Its mass is determined with a high precision of $\sim$0.1% as [2]

$$m_H = 125.18 \pm 0.16\,\text{GeV} = 125.18 \pm 0.14\,(\text{stat}) \pm 0.06\,(\text{syst})\,\text{GeV}. \quad (2.48)$$

The value is the average of a combined measurement of the ATLAS and CMS experiments at $\sqrt{s} = 7$ and 8 TeV [44], and a measurement in the $H \rightarrow ZZ^* \rightarrow l^+l^-l^+l^-$ channel at $\sqrt{s} = 13$ TeV by the CMS experiment [45]. It is not expected that measurements at the LHC can directly observe the theoretical decay width of approximately 4 MeV. An upper limit on the decay width of the Higgs boson at 95% confidence level could be set to 13 MeV by measurements in the WW^* and ZZ^* decay channels [45–48].

Moreover, spin and parity properties were thoroughly studied. By means of the Landau-Yang theorem [49, 50], the observed decay into two photons disproves the hypothesis of a spin-1 particle. Further tests were conducted using hypotheses of CP-even and CP-odd Higgs bosons with spins of up to 2 [38, 51–58]. All results favor a scalar and CP-even configuration, $J^P = 0^+$.

So far, all measurements are compatible with SM predictions and searches for additional Higgs bosons that can only be explained by models beyond the SM were conducted without finding significant evidences. However, the accuracy of analyses in the Higgs sector is still evolving and signs of new physics may be observed in the future.

2.2.3 The Top Quark

The existence of the top quark was suggested in 1973 based on the incompatibility between the observed CP violation and models containing only two generations of quarks [59]. The first experimental observation was reported by the CDF and DØ collaborations at the Tevatron proton-antiproton collider at a center-of-mass energy of $\sqrt{s} = 1.8$ TeV using data corresponding to integrated luminosities of 67 pb^{-1} and 50 pb^{-1}, respectively [60, 61]. The following paragraphs discuss its properties, production processes at proton-proton colliders, and typical decay signatures.

The top quark carries an electrical charge of $Q = +2/3\,e$, a half-integer spin of $1/2$, and is subject to all three fundamental forces described by the SM. With a mass of $m_t = 173.0$ GeV, it is approximately as heavy as a tungsten atom and exhibits the largest Yukawa coupling to the Higgs field among all elementary particles. Therefore, the accurate measurement of Higgs-top coupling is essential for probing predictions of the SM and to constrain various models that reach beyond it. Due to its short lifetime of $\tau_t \approx 5 \cdot 10^{-25}$ s, it decays before hadronization occurs, which gives particles physicists the unique opportunity to study the decay of a bare quark.

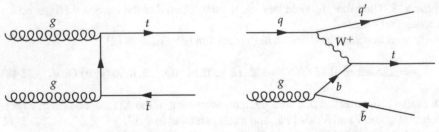

(a) Top quark pair production.   **(b)** Single top quark production (*t*-channel).

Fig. 2.8 Feynman diagrams of two prominent top quark production processes. The diagrams illustrate top quark pair production (**a**), and *t*-channel single top quark production in the four-flavor scheme (4FS) (**b**)

Production Processes

At proton-proton colliders, top quarks are predominantly produced as quark-antiquark pairs in processes involving strong interactions ($t\bar{t}$). A leading-order Feynman diagram of this top quark pair production is shown in Fig. 2.8a. When considering higher orders, the $t\bar{t}$ system is often accompanied by emissions of additional jets. These processes constitute a major background to the $t\bar{t}H$ ($H \rightarrow b\bar{b}$) analysis presented in this thesis and is discussed further in Sect. 2.2.4.

The process in Fig. 2.8a is initiated by two gluons with a *t*-channel top quark propagator. The interaction $gg \rightarrow t\bar{t}$ is in general also possible in the *u*-channel, as well as in the *s*-channel with a gluon propagator. In contrast, $t\bar{t}$ production initiated by quarks, $q\bar{q} \rightarrow t\bar{t}$, is only realizable in the *s*-channel and also requires an anti-quark from the parton sea, which is less probable according to PDFs (cf. Sect. 2.2.1). Therefore, at $\sqrt{s} = 13$ TeV, gluon-initiated processes contribute approximately 90% of the total $t\bar{t}$ production [2]. Considering a top quark mass of $m_t = 172.5$ GeV, the cross section is determined as

$$\sigma_{t\bar{t}} = 831.76 \,^{+19.77}_{-29.20} \,(\text{scale}) \pm 35.06 \,(\text{PDF}+\alpha_s) \,^{+23.18}_{-22.45} \,(\text{mass})\,\text{pb} \qquad (2.49)$$

at NNLO with next-to-next-to-leading-log (NNLL) resummation techniques [62–74]. The uncertainties are estimated by varying the renormalization and factorization scales, $\mu_R = \mu_F = m_t$, independently by factors of $1/2$ and 2 (scale), by a combined treatment of PDF and α_s uncertainties using different PDF sets as recommended by the "PDF4LHC" group [69] (PDF+α_s), and by varying the mass of the top quark by ± 1 GeV in the calculation (mass).

In the second group of processes, a single top quark is produced via weak interactions. Three production channels can be distinguished based on the function of the involved *W* boson. In the *t*-channel process, a *W* boson is exchanged between two quarks in the initial state. One of them is required to be a bottom quark in order to produce a top quark at the *Wb* vertex, which is subject to the CKM matrix element $|V_{tb}| \approx 1$ [2]. In the so-called "five-flavor scheme" (5FS), the probability to obtain a bottom quark from the parton sea is relatively small (Fig. 2.2). By means of the QCD factorization theorem (cf. Sect. 2.2.1), the hard parton interaction can be described using a higher order wherein the bottom quark originates from a gluon in the initial state. This approach is referred to as "four-flavor scheme" (4FS). A t-channel Feynman diagram in the 4FS is exemplarily depicted in Fig. 2.8b. The tW-channel is not inspired by the system of Mandelstam variables. The term tW rather states that the single top quark is produced in association with a W boson, which is realized through a gluon and a bottom quark in the initial state, and either a time-like bottom quark propagator or a space-like top quark propagator. In the s-channel, two quarks in the initial state annihilate into a virtual, time-like W boson propagator, producing a top quark and an accompanying bottom quark. Assuming a top quark mass of $m_t = 172.5\,\text{GeV}$ and renormalization and factorization scales of $\mu_R = \mu_F = m_t$, the net cross section of all production processes amounts to $299.0^{+9.8}_{-8.6}$ pb [69–73, 75–77]. Uncertainties are obtained with approaches similar to those employed for top quark pair production as described above.

Decay Channels

The top quark decays exclusively via weak interactions. In addition to a W boson, the decay almost always involves a bottom quark due to the CKM matrix element of $|V_{tb}| \approx 1$. Other down-type quarks are theoretically allowed but highly disfavored due to corresponding, vanishing off-diagonal values. The implications on top quark pair processes are discussed in the following. Latest results on branching ratio calculations are taken from Ref. [2].

The W boson decays into hadrons with a branching ratio of $BR_{\text{hadr}} = 0.6741 \pm 0.0023$. The most frequent decay channels are $W^+/W^- \rightarrow u\bar{d}/\bar{u}d$ and $W^+/W^- \rightarrow c\bar{s}/\bar{c}s$, which hadronize and lead to jets in the experimental signature. Depending on the momentum of the W boson and the decay angle relative to its motion, the two jets might experimentally not be resolvable as such. The resulting object is often called "fat jet" and studied with subject algorithms [78]. Leptonic decays of the W boson are, within their uncertainties, evenly distributed over the three generations due to lepton flavor universality. The branching ratios for the decays into electrons and muons, including their corresponding neutrinos, are $BR_{e\nu} = 0.1071 \pm 0.0016$ and $BR_{\mu\nu} = 0.1063 \pm 0.0015$, respectively. Decays involving tau leptons are not considered in the analysis presented in this thesis. They decay weakly into further hadrons, e.g. $\tau^- \rightarrow \nu_\tau q\bar{q}'$, or into additional leptons, e.g. $\tau^- \rightarrow \nu_\tau \mu^- \bar{\nu}_\mu$, which are potentially confused with those of the primary W boson decay.

The decay of a pair of top quarks can be characterized by the decays of the subsequent W bosons. Three distinct channels can be distinguished.

full-hadron (FH)   Both W bosons decay into quarks, leading to six jets in the final state. Although the branching ratio is sizable (45.4%), analysis are challenging due to background contributions from multijet processes and combinatorial ambiguities in the reconstruction of the event.

dilepton (DL)   Both W bosons decay into leptons, leading to two jets, two charged leptons, and two neutrinos in the final state. Albeit the combined branching ratio

Table 2.3 Combined branching ratios of two W bosons regarding decays into hadrons and leptonic decays into electrons and muons. Decays involving tau leptons are not considered. Values are taken from Ref. [2]

W decays	Hadronic	Leptonic (e, μ)
Hadronic	BR^2_{hadr}	$BR_{\text{hadr}} \cdot (BR_{e\nu} + BR_{\mu\nu})$
	$45.5 \pm 0.3\%$	$14.4 \pm 0.2\%$
Leptonic (e, μ)	$BR_{\text{hadr}} \cdot (BR_{e\nu} + BR_{\mu\nu})$	$(BR_{e\nu} + BR_{\mu\nu})^2$
	$14.4 \pm 0.2\%$	$4.6 \pm 0.1\%$

is the smallest (10.6%), the signature with two oppositely charged leptons is distinct. Furthermore, the event reconstruction is challenging as the neutrinos are neither detectable nor unambiguously recoverable using the measurement of missing transverse energy and exploiting transverse momentum balance.

single-lepton (SL)   One W boson decays into hadrons while the other one decays into leptons, leading to four jets, one charged lepton, and one neutrino in the final state. Compared to the full-hadron channel, the branching ratio is similar (44.0%) but isolation requirements on the charged lepton can significantly suppress multijet backgrounds. Also, the momentum vector of the single neutrino can be reconstructed with reasonable efficiency using a missing transverse energy measurement in combination with a constraint on the W boson mass in order to reconstruct the full event kinematic.

The branching ratios of the decays of two W bosons are summarized in Table 2.3. The thesis at hand focuses on the single-lepton and dilepton channels with electrons or muons in the final state. Their combined branching ratio amounts to $BR_{t\bar{t},\text{SL+DL}} = 33.3 \pm 0.3\%$.

2.2.4   Signal and Relevant Background Processes

This section discusses characteristics and properties of the $t\bar{t}H$ signal process and relevant background processes in the context of the analysis presented in this thesis. A specific process is considered as a background to an analysis if it cannot be distinctively separated from the signal process by requiring specific thresholds, or "cuts", on measurable observables. This definition also implies that the number of signal events remains on a reasonable level when a set of cuts is applied. The high-dimensional space defined by the domains of kinematic observables is referred to as the "phase space".

Experimental observables are derived from measurements of final state objects, i.e., photons, leptons, and stable hadrons which are clustered into jets. Therefore, relevant background processes can be determined by studying their final states in simulations while varying phase space selection criteria and monitoring their impact on the yield of the signal process. Various effects can cause two processes to appear similar although their nominal final states are different. Objects might be misidentified, undetectable due to fiducial limitations or detector resolution effects, or they emerge from additional emissions. Background processes selected by this means are presented in the second part of this section.

Signal Process

The presented analysis performs a search for $t\bar{t}H$ production and, due to the high branching ratio, focuses on Higgs bosons decaying into a pair of bottom quarks. The bottom quarks hadronize and form b hadrons, whose lifetime of $\tau_0 \approx 1.6 \cdot 10^{-12}$ s is relatively long [2]. This is due to the fact that bottom quarks only decay via electroweak interactions and the decay amplitudes are subject to the small off-diagonal CKM matrix elements V_{cb} and V_{ub} (cf. Sect. 2.1.3) since decays into top quarks are suppressed by the significant mass difference. As a result, b hadrons travel a measurable flight distance before decaying at a secondary vertex, which can be identified by reconstructing and extrapolating trajectories of the decay products. Typical distances at a momentum of $p_b = 100$ GeV are $d = c\tau_0 p_b/m_b \approx 1$ cm. In case a jet is identified to originate from a b hadron decay, it is attributed a so-called "b tag" (cf. Sect. 3.3.6).

Under the assumption that the Higgs boson decay occurs perpendicular to the direction of its motion, the spatial angle between the two jets in the observer's reference frame at typical momenta of $p_H = 100$ GeV amounts to $\phi \approx 103°$. Therefore, one can estimate that in the majority of cases the two jets exhibit a sufficiently large spatial separation, allowing for their resolved measurement and identification based on features of displaced secondary vertices.

Signal processes in which the Higgs boson decays into particles other than a pair of bottom quarks, such as $H \to W^+W^-$ and $H \to \tau^+\tau^-$, are taken into account in the following. Despite their minor expected yield due to smaller branching ratios and different final-state signature, events of those processes can potentially pass phase space selection criteria and contribute to the total number of $t\bar{t}H$ signal events.

Moreover, the decay of the $t\bar{t}$ system is considered in the single-lepton and dilepton decay channels (cf. Sect. 2.2.3). A corresponding leading-order Feynman diagram is presented in Fig. 2.9a. It should be noted that more diagrams exist to describe $t\bar{t}H$ production. An example is the production of a pair of top quarks (cf. Sect. 2.2.3) where one top quark emits a Higgs boson. Similarly to generic $t\bar{t}$ production, gluon-initiated processes have the largest contribution to the total $t\bar{t}H$ cross section at $\sqrt{s} = 13$ TeV [2].

In total, the measurable final state consists of six jets and an isolated lepton in the single-lepton, and four jets and two isolated leptons with opposite charge in the dilepton channel, respectively. In both cases, four jets are supposed to originate from b-hadron decays and a significant amount of missing transverse energy is expected due to the non-detectable neutrinos. Given the high combinatorial complexity due the number of jets and the typical detector resolution of jet observables, the full reconstruction of the event is rather challenging. The net cross section is

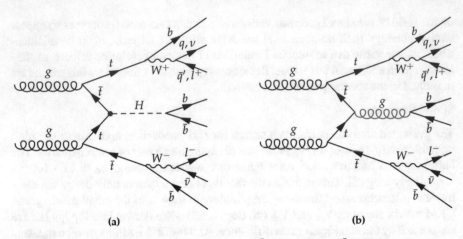

Fig. 2.9 Feynman diagrams showing $t\bar{t}H$ ($H \to b\bar{b}$) (**a**) and $t\bar{t}+b\bar{b}$ production (**b**) in the single-lepton and dilepton $t\bar{t}$ decay channels. Their final state is identical and despite different spin and color charge relations, the event topology is quite similar. It should be noted that more possible diagrams exist for both processes

$$\sigma_{t\bar{t}H,b\bar{b},\mathrm{SL+DL}} = \sigma_{t\bar{t}H} \cdot BR_{H\to b\bar{b}} \cdot BR_{t\bar{t},\mathrm{SL+DL}} = 98.4\,^{+6.9}_{-9.9}\,\mathrm{fb}, \qquad (2.50)$$

which corresponds to $\sim$3500 produced events in the dataset recorded by the CMS detector in 2016 with an integrated luminosity of 35.9 fb^{-1}.

Background Processes

The most dominant background contributions to the signal process discussed above arise from top quark pair production (cf. Sect. 2.2.3). In comparison, its inclusive cross section, calculated at NNLO with NNLL resummation [62–74], is larger by a factor of $\sim$2800 given the same final states of the $t\bar{t}$ system. Especially in the event of additionally emitted, heavy-flavor jets ($t\bar{t}+$hf), i.e., jets originating from bottom or charm quarks, the final states are very similar. Fig. 2.9b exemplarily shows a Feynman diagram of top quark pair production with an additional bottom quark pair induced by the splitting of a gluon. Despite differences caused by the spin and color-charge of the gluon, the process is very similar to the $t\bar{t}H$ ($H \to b\bar{b}$) signal process in terms of final-state content and kinematic topology.

The theoretical description of the production mechanisms of these $t\bar{t}+$hf contributions is ambiguous. For example, processes with additional bottom quarks can be modeled via $gg \to t\bar{t}g$ matrix element calculations in (N)LO with collinear $g \to b\bar{b}$ splittings in the parton shower, $gb \to t\bar{t}b$ production using 4FS and 5FS calculations, or solely via collinear gluon splittings in the parton shower [79–82]. Currently neither higher-order theoretical calculations nor experimental methods can constrain expected rates with an accuracy better than 35% [79–81, 83].

However, a clear and robust process definition is required in the scope of this analysis for two particular reasons. First, production processes might be subject to different theoretical uncertainties, which must be properly assigned to establish a reasonable modeling of recorded collision data with simulated events. Second, common analysis strategies are based on the separation of signal and background contributions to define specific signal- and background-enriched phase space regions in which the measurement is performed. In the presented analysis, the most relevant background contributions are also separated from each other (cf. Sections 4 and 7). In case of inadequate process definitions, the efficiency of this separation approach is impaired, which in turn reduces the measurement sensitivity.

The definition of the particular $t\bar{t}$ subprocesses is described in the following. After the simulation of events using matrix element generators, parton showering, and phenomenological hadronization processes (cf. Sect. 2.2.1), final state particles are clustered into jets using the anti-k_T algorithm with a distance parameter of $\Delta R = 0.4$. If a jet can be associated to the hard scattering process, i.e., if it originates from the top quark or hadronic W boson decays, or if it exhibits a transverse momentum below 20 GeV, it is rejected in subsequent considerations. The clustering history of the remaining jets is traversed backwards up to the hadronization stage and contains, in particular, the decays of unstable hadrons. Based on this, five $t\bar{t}$ subprocesses are defined by counting the numbers of additional jets and contained b and c hadrons, whereby hadron flavor information is determined using the "ghost association" method [84, 85].

- $t\bar{t}+b\bar{b}$: The event has at least two additional jets, which each contain at least one b hadron.
- $t\bar{t}+b/\bar{b}$: The event has one additional jet, which contains exactly one b hadron.
- $t\bar{t}+2b$: The event has one additional jet, which contains at least two distinct b hadrons.
- $t\bar{t}+c\bar{c}$: None of the above criteria applies and the event has at least one additional jet, which contains at least one c hadron.
- $t\bar{t}+$lf: None of the above criteria applies for the event.

In the following, these five processes are also referred to as $t\bar{t}+$X, while the first four describe events with additional heavy-flavor jets, $t\bar{t}+$hf. Since the algorithm is based on final-state particles, i.e., particles after hadronization, the $t\bar{t}+$X process definitions themselves can be seen as phenomenological. In particular, they depend on the choice of the matrix element generator and parton shower algorithm, which must be taken into account by considering appropriate systematic uncertainties in the statistical treatment.

Besides the five $t\bar{t}+$X processes discussed above, four additional groups of processes are included in the description of the overall expected background. Although their nominal final state signatures are relatively different, additional emissions of jets, object misidentification, and comparably large cross sections necessitate their consideration. They are discussed in the following within groups that are employed throughout the thesis at hand. A summary of their cross section is presented in Table 2.4. If not stated otherwise, cross sections are calculated at $\sqrt{s} = 13$ TeV, assuming $m_t = 172.5$ GeV, $m_W = 80.398$ GeV, and $m_Z = 91.1876$ GeV.

Table 2.4 Signal and background cross sections at $\sqrt{s} = 13\,\text{TeV}$ that are used throughout this analysis. The dilepton mass m_{ll} and the sum of transverse jet momenta H_T are given in GeV. Uncertainties denoted by "PDF" include α_s variations. Cross sections involving top quarks assume renormalization and factorization scales of $\mu_R = \mu_F = m_t = 172.5\,\text{GeV}$

Process	Channel/phase space		Cross section	References
$t\bar{t}H$	$H \to b\bar{b}$		$295.3\,^{+5.8\%}_{-9.2\%}$ (scale) $\pm 3.6\%$ (PDF) fb	[2, 30]
	$H \not\to b\bar{b}$		$211.8\,^{+5.8\%}_{-9.2\%}$ (scale) $\pm 3.6\%$ (PDF) fb	
$t\bar{t}$	Inclusive		$831.76\,^{+19.77}_{-29.20}$ (scale) ± 35.06 (PDF) pb	[62–74]
Single $t/\bar{t}$	t-channel	t	$136.02\,^{+4.09}_{-2.92}$ (scale) ± 3.52 (PDF) pb	[69–73, 75, 76]
		$\bar{t}$	$80.95\,^{+2.53}_{-1.71}$ (scale) ± 3.18 (PDF) pb	
	tW-channel	$t/\bar{t}$	35.85 ± 0.90 (scale) ± 1.70 (PDF) pb	[77]
	s-channel	t	$6.35\,^{+0.18}_{-0.15}$ (scale) ± 0.14 (PDF) pb	[69–73, 75, 76]
		$\bar{t}$	$3.97\,^{+0.11}_{-0.09}$ (scale) ± 0.15 (PDF) pb	
W	$W \to l\nu$	$70 < H_T \leq 100$	1319 pb	[22, 86–89]
		$100 < H_T \leq 200$	1345 pb	
		$200 < H_T \leq 400$	359.7 pb	
		$400 < H_T \leq 600$	48.91 pb	
		$600 < H_T \leq 800$	12.05 pb	
		$800 < H_T \leq 1200$	5.501 pb	
		$1200 < H_T \leq 2500$	1.329 pb	
		$H_T > 2500$	32.16 fb	
Z/γ^*	$Z/\gamma^* \to ll$ $5 < m_{ll} \leq 50$	$70 < H_T \leq 100$	302.2 pb	[22, 86–89]
		$100 < H_T \leq 200$	224.2 pb	
		$200 < H_T \leq 400$	37.20 pb	
		$400 < H_T \leq 600$	3.581 pb	
		$H_T > 600$	1.124 pb	
	$Z/\gamma^* \to ll$ $m_{ll} > 50$	$70 < H_T \leq 100$	175.3 pb	[22, 86–89]
		$100 < H_T \leq 200$	147.4 pb	
		$200 < H_T \leq 400$	40.99 pb	
		$400 < H_T \leq 600$	5.678 pb	
		$600 < H_T \leq 800$	1.367 pb	
		$800 < H_T \leq 1200$	630.4 fb	
		$1200 < H_T \leq 2500$	151.4 fb	
		$H_T > 2500$	3.565 fb	
WW	inclusive		$118.7\,^{+2.5\%}_{-2.2\%}$ (scale) pb	[90]
WZ	inclusive		47.12 ± 1.78 (scale) ± 2.24 (PDF) pb	[91, 92]
ZZ	inclusive		31.73 ± 1.02 (scale) ± 1.19 (PDF) pb	[91, 92]
$t\bar{t}W$	$W \to l\nu$		204.3 fb	[22, 86–89]
	$W \to q\bar{q}'$		406.2 fb	
$t\bar{t}Z$	$Z \to ll/\nu\nu$		252.9 fb	[22, 86–89]
	$Z \to q\bar{q}$		529.7 fb	

- Single $t/\bar{t}$: Production of a single top (anti)quark as introduced in Sect. 2.2.3. The t- and s-channel cross sections are calculated in NLQ QCD [69–73, 75, 76], whereas the tW-channel is known up to NNLO [77].
- V+jets: Processes involving the production of a single W or Z/γ^* vector boson and additional jet emissions. These processes can resemble the signal process especially in the event of decays into (charged) leptons. To ensure a more precise background composition, different phase space regions are modeled separately using different intervals of the scalar sum of transverse momenta H_T, and, in addition for Z/γ^* bosons, using the combined dilepton mass m_{ll}. The cross sections stated in Table 2.4 are calculated in LO using the MADGRAPH5_aMC@NLO program [22, 86–89].
- Diboson: Production of a pair of vector bosons, i.e., WW, WZ, and ZZ. The cross section for WW production is obtained at NNLO QCD using a fixed renormalization and factorization scale corresponding to m_W [90]. WZ and ZZ processes are calculated in NLO using the MCFM (v6.6) program at scales of $(m_W + m_Z)/2$ and m_Z, respectively [91, 92].
- $t\bar{t}+V$: Production of a top quark pair in association with a W or Z vector boson. Although these processes are rare, especially hadronic decays of the additional vector boson lead to a final state that is very similar to the signal process. Cross sections are calculated with NLO accuracy using the MADGRAPH5_aMC@NLO program [22, 86–89].

2.2.5 Results from Previous Searches

This section summarizes the results of previous searches for $t\bar{t}H$ production with $H \rightarrow b\bar{b}$ decays in the single-lepton and dilepton $t\bar{t}$ decay channels. Various analyses were performed by the CMS and ATLAS experiments using proton-proton collisions provided by the LHC at multiple center-of-mass energies. Table 2.5 presents upper limits at 95% confidence level, measured $t\bar{t}H$ signal strength modifiers, and corresponding significances describing the excess over the background expectation in units of standard deviations. It should be noted that the analysis in the single-lepton channel, which dominates the overall $t\bar{t}H$ ($H \rightarrow b\bar{b}$) sensitivity, as presented in this thesis is part of the stated result of the CMS experiment at $\sqrt{s} = 13\,\text{TeV}$ [93].

All results are compatible with expectations derived from the SM. To date, the most sensitive search is performed by the CMS experiment at $\sqrt{s} = 13\,\text{TeV}$. Given recorded collision data corresponding to an integrated luminosity of $35.9\,\text{fb}^{-1}$, the background-only hypothesis, i.e., the absence of $t\bar{t}H$ production, is excluded with a confidence better than 95% (expected limit). The signal strength modifier, $\mu = \sigma/\sigma_{\text{SM}}$, of 0.72 ± 0.45 is in reasonable agreement with the SM expectation within experimental, theoretical, and statistical uncertainties. The significance over the expected background contribution is observed to be 1.6σ, while a value of 2.2σ is expected from simulations.

Table 2.5 Previous and new results of searches for $t\bar{t}H$ ($H \to b\bar{b}$) production in the single-lepton and dilepton $t\bar{t}$ decay channels by the ATLAS and CMS collaborations. Upper limits correspond to 95% confidence levels. The significance describes the measured excess over the expected background in units of standard deviations. The result of the CMS experiment at $\sqrt{s} = 13\,\text{TeV}$ is based on the analysis in the single-lepton channel as presented in this thesis [93]

	$\sqrt{s}$/TeV	$\mathcal{L}/fb^{-1}$	Limit (expected)	σ/σ_{SM}	Significance (expected)	Strategy	Reference
ATLAS	8	20.3	3.4 (2.2)	1.5 ± 1.1	1.4 (1.1)	NN+MEM	CERN-EP-2015-047 [95]
	13	36.1	2.0 (1.2)	$0.84^{+0.64}_{-0.61}$	1.4 (1.6)	BDT+MEM	CERN-EP-2017-291 [94]
CMS	7, 8	5.0, 5.1	5.8 (5.2)	–	–	BDT	CERN-EP-2013-027 [96]
	7, 8	5.0, 19.3	4.1 (3.5)	0.7 ± 1.9	–	BDT	CERN-EP-2014-189 [97]
	8	19.5	4.2 (3.3)	$1.2^{+1.6}_{-1.5}$	–	MEM	CERN-EP-2015-016 [98]
	13	35.9	1.5 (0.9)	0.72 ± 0.45	1.6 (2.2)	DNN+BDT+MEM	CERN-EP-2018-065 [93]

In addition, the cross sections for $\sqrt{s} = 8$ and 13 TeV are visualized in Fig. 2.10 along with the theoretical prediction and its uncertainty entailing variations of renormalization and factorization scales, particle density functions, and α_s values [30]. The cross sections are extracted by scaling the signal strength modifiers with the SM prediction at the respective center-of-mass energies. None of the listed searches initially constrain the signal strength modifier to positive values and therefore, transformed cross sections and their uncertainties can overlap with unphysical negative values. With increasing measurement significance this circumstance is expected to vanish.

At $\sqrt{s} = 13\,\text{TeV}$, the measurement uncertainties of both experiments are dominated by contributions from systematic effects. In the analysis performed by the CMS collaboration, they account for an uncertainty of ± 0.24 on the total uncertainty of ± 0.45 on the measured signal strength modifier [93]. The ATLAS collaboration measures a contribution of $+0.57/-0.54$ on a total uncertainty of $+0.64/-0.61$ [94]. As discussed in Sect. 2.2.4, this is mainly driven by the uncertain prediction of $t\bar{t}$ production in association with additional heavy-flavor jets. In order to cope with this challenge, various analysis strategies involving sophisticated machine learning algorithms were employed. The measurement strategy of the presented analysis is discussed in Sect. 4.

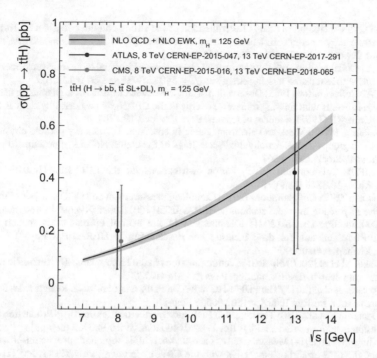

Fig. 2.10 Illustration of the $t\bar{t}H$ cross section as predicted by theoretical calculations [30], and measurements performed by the CMS [93, 98] and ATLAS experiments [94, 95]. Theory uncertainties entail renormalization and factorization scales, PDF, and α_s variations. The measured cross section values emerge from the transformation of the signal strength modifier $\mu = \sigma/\sigma_{SM}$, which does not follow any prior constraint in the fitting procedure. Therefore, in case of statistical underfluctuations in data, the measured $[+1\,\sigma, -1\,\sigma]$ interval might have a negative component, as visible in the CMS measurement at 8 TeV

References

1. 't Hooft G, Veltman M, (1972) Regularization and renormalization of gauge fields. Nucl Phys B 44:189
2. Tanabashi M et al (2018) (Particle Data Group), The review of particle physics. Phys Rev D 98:030001
3. LHCb Collaboration (2017) Observation of $J/\psi\,\phi$ structures consistent with exotic states from amplitude analysis of $B^+ \rightarrow J/\psi\,\phi\,K^+$ decays. Phys Rev Lett 118:022003. arXiv:1606.07895 [hep-ex]
4. LHCb Collaboration (2015) Observation of $J/\psi\,\rho$ resonances consistent with pentaquark states in $\Lambda_b^0 \rightarrow J/\psi\,K^-\,\rho$. Phys Rev Lett 115:072001. arXiv:1507.03414 [hep-ex]
5. Griffiths D (2008) Introduction to elementary particles, vol 2. Wiley Vch, New Jersey
6. Altarelli G (2017) Collider physics within the standard model: a primer. Springer, Berlin. arXiv:1303.2842 [hep-ph]
7. Altarelli G (2005) The standard model of particle physics. arXiv:hep-ph/0510281
8. CMS Collaboration (2017) Measurement and QCD analysis of double-differential inclusive jet cross sections in pp collisions at $\sqrt{s} = 8$ TeV and ratios to 2.76 and 7 TeV. JHEP 03:156. arXiv:1609.05331 [hep-ex]

9. ATLAS Collaboration (2017) Determination of the strong coupling constant α_s from transverse energy–energy correlations in multijet events at $\sqrt{s}$ =8 TeV using the ATLAS detector. Eur Phys J C 77:872. arXiv:1707.02562 [hep-ex]
10. UA1 Collaboration (1983) Experimental observation of isolated large transverse energy electrons with associated missing energy at $\sqrt{s}$ = 540 GeV. Phys Lett B 122:103
11. UA2 Collaboration (1983) Observation of single isolated electrons of high transverse momentum in events with missing transverse energy at the CERN $p\bar{p}$ collider. Phys Lett B 122:476
12. Weinberg S (1967) A model of leptons. Phys Rev Lett 19:1264
13. Salam A (1968) Weak and electromagnetic interactions. Elementary particle physics: relativistic groups and analyticity. In: Proceedings of the eighth Nobel symposium. Almqvist & Wiksell, Stockholm, p 367
14. NNPDF Collaboration (2015) Parton distributions for the LHC Run II. JHEP 04:040. arXiv:1410.8849 [hep-ph]
15. H1 and ZEUS Collaborations (2010) Combined measurement and QCD analysis of the inclusive $e^{\pm}p$ scattering cross sections at HERA. JHEP 1001:109. arXiv:0911.0884 [hep-ex]
16. Collaboration ZEUS (2001) Measurement of the neutral current cross section and F_2 structure function for deep inelastic $e + p$ scattering at HERA. Eur Phys J C 21:443 arXiv:hep-ex/0105090
17. Vakili M et al (2000) Nuclear structure functions in the large x large Q^2 kinematic region in neutrino deep inelastic scattering. Phys Rev D 61:052003
18. Accardi A et al (2015) The PDF4LHC report on PDFs and LHC data: Results from Run I and preparation for Run II. arXiv:1507.00556 [hep-ph]
19. Demartin F et al (2010) The impact of PDF and α_s uncertainties on Higgs Production in gluon fusion at hadron colliders. Phys Rev D 82:014002. arXiv:1004.0962 [hep-ph]
20. Klingebiel D (2014) Measurement of the t-channel single-top-quark-production cross section and the CKM-matrix element V_{tb} with the CMS experiment. PhD Thesis, RWTH Aachen University. http://publications.rwth-aachen.de/record/444972
21. Dobbs MA et al (2004) Les Houches guidebook to Monte Carlo generators for hadron collider physics. In: Proceedings of the 3rd Les Houches workshop C03-05-25.5, p 411. arXiv:hep-ph/0403045
22. Frixione S, Webber BR (2002) Matching NLO QCD computations and parton shower simulations. JHEP 06:029 arXiv:hep-ph/0204244
23. Frixione S, Nason P, Oleari C (2007) Matching NLO QCD computations with parton shower simulations: the POWHEG method. JHEP 11:070. arXiv:0709.2092 [hep-ph]
24. Andersson B, Gustafson G, Ingelmann G, Sjöstrand T (1983) Parton fragmentation and string dynamics. Phys Rep 97:31
25. Catani S, Dokshitzer YL, Seymour MH, Webber BR (1993) Longitudinally invariant K_t clustering algorithms for hadron hadron collisions. Nucl Phys B 406:187
26. Cacciari M, Salam GP, Soyez G (2008) The anti-k_t jet clustering algorithm. JHEP 04:063. arXiv:0802.1189 [hep-ph]
27. CMS Collaboration (2009) A Cambridge-Aachen (C-A) based jet algorithm for boosted top-jet tagging. In: CMS physics analysis summary CMS-PAS-JME-09-001. http://cds.cern.ch/record/1194489
28. Cacciari M, Salam GP, Soyez G (2012) FastJet user manual. Eur Phys J C 72:1896. arXiv:1111.6097 [hep-ph]
29. Gleisberg T et al (2009) Event generation with SHERPA 1.1. JHEP 02:007. arXiv:0811.5622 [hep-ph]
30. LHC Higgs Cross Section Working Group (2017) Handbook of LHC Higgs cross sections: 4. Deciphering the nature of the Higgs sector. CERN yellow reports: monographs. arXiv:1610.07922 [hep-ph]. http://cds.cern.ch/record/2227475
31. CMS Collaboration (2012) Observation of a new boson at a mass of 125 GeV with the CMS experiment at the LHC. Phys Lett B 716:30. arXiv:1207.7235 [hep-ex]
32. ATLAS Collaboration (2012) Observation of a new particle in the search for the standard model Higgs boson with the ATLAS detector at the LHC. Phys Lett B 716:1. arXiv:1207.7214 [hep-ex]

33. ATLAS Collaboration (2018) Observation of $H \rightarrow b\bar{b}$ decays and VH production with the ATLAS detector. Phys Lett B 786:59. arXiv:1808.08238 [hep-ex]
34. CMS Collaboration (2018) Observation of $t\bar{t}H$ production. Phys Rev Lett 120(23):231801. arXiv:1804.02610 [hep-ex]
35. ATLAS Collaboration (2018) Observation of Higgs boson production in association with a top quark pair at the LHC with the ATLAS detector. Phys Lett B 785:173. arXiv:1806.00425 [hep-ex]
36. CMS Collaboration (2014) Measurement of Higgs boson production and properties in the WW decay channel with leptonic final states. JHEP 01:096. arXiv:1312.1129 [hep-ex]
37. ATLAS Collaboration (2015) Observation and measurement of Higgs boson decays to WW^* with the ATLAS detector. Phys Rev D 92:012006. arXiv:1412.2641 [hep-ex]
38. ATLAS Collaboration (2018) Measurement of the Higgs boson coupling properties in the $H \rightarrow ZZ^* \rightarrow 4l$ decay channel at $\sqrt{s} = 13$ TeV with the ATLAS detector. JHEP 03:095. arXiv:1712.02304 [hep-ex]
39. CMS Collaboration (2014) Observation of the diphoton decay of the Higgs boson and measurement of its properties. Eur Phys J C 74:3076. arXiv:1407.0588 [hep-ex]
40. ATLAS Collaboration (2014) Measurement of Higgs boson production in the diphoton decay channel in pp collisions at center-of-mass energies of 7 and 8 TeV with the ATLAS detector. Phys Rev D 90:112015. arXiv:1408.7084 [hep-ex]
41. CMS Collaboration (2017) Observation of the Higgs boson decay to a pair of τ leptons. Phys Lett B 779:283. arXiv:1708.00373 [hep-ex]
42. CMS Collaboration (2018) Observation of Higgs boson decay to bottom quarks. Phys Rev Lett 121:121801. arXiv:1808.08242 [hep-ex]
43. CMS Collaboration (2019) Combined measurements of the Higgs boson's couplings at $\sqrt{s} = 13$ TeV. Eur Phys J C 79:421. arXiv:1809.10733 [hep-ex]
44. CMS and ATLAS Collaborations (2016) Measurements of the Higgs boson production and decay rates and constraints on its couplings from a combined ATLAS and CMS analysis of the LHC pp collision data at $\sqrt{s} = 7$ and 8 TeV. JHEP 08:045. arXiv:1606.02266 [hep-ex]
45. CMS Collaboration (2017) Measurements of properties of the Higgs boson decaying into the four-lepton final state in pp collisions at $\sqrt{s} = 13$ TeV. JHEP 11:047. arXiv:1706.09936 [hep-ex]
46. ATLAS Collaboration (2015) Constraints on the off-shell Higgs boson signal strength in the high-mass ZZ and WW final states with the ATLAS detector. Eur Phys J C 75:335. arXiv:1503.01060 [hep-ex]
47. CMS Collaboration (2015) Limits on the Higgs boson lifetime and width from its decay to four charged leptons. Phys Rev D 92:072010. arXiv:1507.06656 [hep-ex]
48. CMS Collaboration (2016) Search for Higgs boson off-shell production in proton-proton collisions at 7 and 8 TeV and derivation of constraints on its total decay width. JHEP 09:051. arXiv:1605.02329 [hep-ex]
49. Landau LD (1948) On the angular momentum of a system of two photons. Dokl Akad Nauk Ser Fiz 60:107
50. Yang C (1950) Selection rules for the dematerialization of a particle into two photons. Phys Rev 77:242
51. CMS Collaboration (2013) On the mass and spin-parity of the Higgs boson candidate via its decays to Z boson pairs. Phys Rev Lett 110:081803. arXiv:1212.6639 [hep-ex]
52. CMS Collaboration (2015) Constraints on the spin-parity and anomalous HVV couplings of the Higgs boson in proton collisions at 7 and 8 TeV. Phys Rev D 92:012004. arXiv:1411.3441 [hep-ex]
53. CMS Collaboration (2016) Combined search for anomalous pseudoscalar HVV couplings in $VH(H \rightarrow b\bar{b})$ production and $H \rightarrow VV$ decay. Phys Lett B 759:672. arXiv:1602.04305 [hep-ex]
54. CMS Collaboration (2014) Measurement of the properties of a Higgs boson in the four-lepton final state. Phys Rev D 89:092007. arXiv:1312.5353 [hep-ex]

55. ATLAS Collaboration (2013) Evidence for the spin-0 nature of the Higgs boson using ATLAS data. Phys Lett B 726:120. arXiv:1307.1432 [hep-ex]
56. ATLAS Collaboration (2015) Study of the spin and parity of the Higgs boson in diboson decays with the ATLAS detector. Eur Phys J C 75:476. arXiv:1506.05669 [hep-ex]
57. ATLAS Collaboration (2015) Determination of spin and parity of the Higgs boson in the $WW^* \rightarrow e\nu\mu\nu$ decay channel with the ATLAS detector. Eur Phys J C 75:231. arXiv:1503.03643 [hep-ex]
58. ATLAS Collaboration (2016) Test of CP invariance in vector-boson fusion production of the Higgs boson using the Optimal Observable method in the ditau decay channel with the ATLAS detector. Eur Phys J C 76:658. arXiv:1602.04516 [hep-ex]
59. Kobayashi M, Maskawa T (1973) CP violation in the renormalizable theory of weak interaction. Prog Theor Phys 49:652
60. Collaboration CDF (1995) Observation of top quark production in $\bar{p}p$ collisions. Phys Rev Lett 74:2626 arXiv:hep-ex/9503002
61. Collaboration DØ (1995) Observation of the top quark. Phys Rev Lett 74:2632 arXiv:hep-ex/9503003
62. Beneke M et al (2012) Hadronic top-quark pair production with NNLL threshold resummation. Nucl Phys B 855:695. arXiv:1109.1536 [hep-ph]
63. Cacciari M et al (2012) Top-pair production at hadron colliders with next-to-next-to-leading logarithmic soft-gluon resummation. Phys Lett B 710:612. arXiv:1111.5869 [hep-ph]
64. Bärnreuther P, Czakon M, Mitov A (2012) Percent level precision physics at the tevatron: first genuine NNLO QCD corrections to $q\bar{q} \rightarrow t\bar{t}$. Phys Rev Lett 109:132001. arXiv:1204.5201 [hep-ph]
65. Czakon M, Mitov A (2012) NNLO corrections to top-pair production at hadron colliders: the all-fermionic scattering channels. JHEP 1212:054. arXiv:1207.0236 [hep-ph]
66. Czakon M, Mitov A (2013) NNLO corrections to top pair production at hadron colliders: the quark-gluon reaction. JHEP 1301:080. arXiv:1210.6832 [hep-ph]
67. Czakon M, Fiedler P, Mitov A (2013) The total top quark pair production cross-section at hadron colliders through $O(\alpha_s^4)$. Phys Rev Lett 110:252004, http://arxiv.org/abs/1303.6254 arXiv:1303.6254
68. Czakon M, Mitov A (2014) Top++: a program for the calculation of the top-pair cross-section at hadron colliders. Comput Phys Commun 185:2930. arXiv:1112.5675 [hep-ph]
69. Botje M et al (2011) The PDF4LHC working group interim recommendations. arXiv:1101.0538 [hep-ph]
70. Martin A et al (2009) Parton distributions for the LHC. Eur Phys J C 63:189. arXiv:0901.0002 [hep-ph]
71. Martin A et al (2009) Uncertainties on α_s in global PDF analyses and implications for predicted hadronic cross sections. Eur Phys J C 64:653. arXiv:0905.3531 [hep-ph]
72. Lai H-L et al (2010) New parton distributions for collider physics. Phys Rev D 82:074024. arXiv:1007.2241 [hep-ph]
73. Ball RD et al (2013) Parton distributions with LHC data. Nucl Phys B 867:244. arXiv:1207.1303 [hep-ph]
74. Gao J et al (2014) CT10 next-to-next-to-leading order global analysis of QCD. Phys Rev D 89(3):03009. arXiv:1302.6246 [hep-ph]
75. Aliev M et al (2011) HATHOR - HAdronic Top and Heavy quarks crOss section calculator. Comput Phys Commun 182:1034. arXiv:1007.1327 [hep-ph]
76. Kant P et al (2015) HATHOR for single top-quark production: updated predictions and uncertainty estimates for single top-quark production in hadronic collisions. Comput Phys Commun 191:74. arXiv:1406.4403 [hep-ph]
77. Kidonakis N (2010) Two-loop soft anomalous dimensions for single top quark associated production with a W^- or H^-. Phys Rev D 82:054018. arXiv:1005.4451 [hep-ph]
78. Salam GP (2010) Towards jetography. Eur Phys J C 67:637. arXiv:0906.1833 [hep-ph]
79. Ježo T, Lindert JM, Moretti N, Pozzorini S (2018) New NLOPS predictions for $t\bar{t} + b$-jet production at the LHC. Eur Phys J C 78:502. arXiv:1802.00426 [hep-ph]

80. Garzelli MV, Kardos A, Trocsanyi Z (2015) Hadroproduction of $t\bar{t}b\bar{b}$ final states at LHC: predictions at NLO accuracy matched with Parton Shower. JHEP 03:083. arXiv:1408.0266 [hep-ph]

81. Bevilacqua G, Garzelli M, Kardos A (2017) $t\bar{t}b\bar{b}$ hadroproduction with massive bottom quarks with PowHel. arXiv:1709.06915 [hep-ph]

82. Cascioli F et al (2014) NLO matching for ttbb production with massive b-quarks. Phys Lett B 734:210. arXiv:1309.5912 [hep-ph]

83. CMS Collaboration (2018) Measurements of $t\bar{t}$ cross sections in association with b jets and inclusive jets and their ratio using dilepton final states in pp collisions at $\sqrt{s}$ = 13 TeV. Phys Lett B 776:335. arXiv:1705.10141 [hep-ex]

84. CMS Collaboration (2018) Identification of heavy-flavour jets with the CMS detector in pp collisions at 13 TeV. JINST 13:P05011. arXiv:1712.07158 [physics.ins-det]

85. Cacciari M, Salam GP (2008) Pileup subtraction using jet areas. Phys Lett B 659:119. arXiv:0707.1378 [hep-ph]

86. Alwall J et al (2014) The automated computation of tree-level and next-to-leading order differential cross sections, and their matching to parton shower simulations. JHEP 07:079. arXiv:1405.0301 [hep-ph]

87. Frederix R et al (2018) The automation of next-to-leading order electroweak calculations. JHEP 07:185. arXiv:1804.10017 [hep-ph]

88. Frixione S, Laenen E, Motylinski P, Webber BR (2007) Angular correlations of lepton pairs from vector boson and top quark decays in Monte Carlo simulations. JHEP 04:081 arXiv:hep-ph/0702198

89. Artoisenet P, Frederix R, Mattelaer O, Rietkerk R (2013) Automatic spin-entangled decays of heavy resonances in Monte Carlo simulations. JHEP 03:015. arXiv:1212.3460 [hep-ph]

90. Gehrmann T et al (2014) W^+W^- production at hadron colliders in NNLO QCD. Phys Rev Lett 113:212001. arXiv:1408.5243 [hep-ph]

91. Campbell JM, Ellis RK (1999) An update on vector boson pair production at hadron colliders. Phys Rev D 60:113006 arXiv:hep-ph/9905386

92. Campbell JM, Ellis RK (2010) MCFM for the tevatron and the LHC. Nucl Phys Proc Suppl 205-206:10. arXiv:1007.3492 [hep-ph]

93. CMS Collaboration (2019) Search for $t\bar{t}H$ production in the $H \rightarrow b\bar{b}$ decay channel with leptonic $t\bar{t}$ decays in proton-proton collisions at $\sqrt{s}$ = 13 TeV. JHEP 03:026. arXiv:1804.03682 [hep-ex]

94. ATLAS Collaboration (2018) Search for the standard model Higgs boson produced in association with top quarks and decaying into a $b\bar{b}$ pair in pp collisions at $\sqrt{s}$ = 13 TeV with the ATLAS detector. Phys Rev D 97:072016. arXiv:1712.08895 [hep-ex]

95. ATLAS Collaboration (2015) Search for the standard model Higgs boson produced in association with top quarks and decaying into $b\bar{b}$ in pp collisions at $\sqrt{s}$ = 8 TeV with the ATLAS detector. Eur Phys J C 75:349. arXiv:1503.05066 [hep-ex]

96. CMS Collaboration (2013) Search for the standard model Higgs boson produced in association with a top-quark pair in pp collisions at the LHC. JHEP 05:145. arXiv:1303.0763 [hep-ex]

97. CMS Collaboration (2014) Search for the associated production of the Higgs boson with a top-quark pair. JHEP 09:087. arXiv:1408.1682 [hep-ex]

98. CMS Collaboration (2015) Search for a standard model Higgs boson produced in association with a top-quark pair and decaying to bottom quarks using a matrix element method. Eur Phys J C 75:251. arXiv:1502.02485 [hep-ex]

Chapter 3
Experimental Setup

This section describes the experimental environment in which this analysis is performed. It comprises the Large Hadron Collider (LHC), which provides proton-proton collision events at high energies, the Compact Muon Solenoid (CMS) detector at one of its interaction points to study induced interaction processes, as well as software and algorithms to reconstruct physics objects based on detector measurements. These fundamental techniques are the product of decades of research and development pursued by numerous groups deeply involved in high energy physics experiments. Although each topic independently constitutes an extensive field of research, they are discussed in brevity in the following paragraphs.

3.1 The Large Hadron Collider

The LHC is the largest and most powerful particle accelerator, storage ring, and collider ever constructed, and is operated by the European Organization for Nuclear Research (CERN). It is located in a circular tunnel, formerly built for the Large Electron-Positron Collider (LEP) [1], with a circumference of 26.7 km at a depth between 45 and 170 m under ground close to Geneva at the Franco-Swiss border. After 10 years of construction, it was commissioned in 2008 and first particle collisions were induced in 2009. Its working principles, design parameters, and implications on associated experiments are presented in the subsequent paragraphs. They loosely follow Refs. [2–5], which discuss the introduced concepts in greater detail.

Within two parallel, evacuated beam pipes, the LHC is designed to accelerate protons to energies up to 7 TeV, and lead atoms up to 2.76 TeV per nucleon, respectively. The following description refers to protons if not stated otherwise. Before being injected into the LHC, particles are pre-accelerated in a chain of upstream beam facilities [4]. An illustration of the accelerator complex at CERN is shown in

© The Author(s), under exclusive license to Springer Nature Switzerland AG 2021   41
M. Rieger, *Search for tt̄H Production in the H → bb̄ Decay Channel*, Springer Theses,
https://doi.org/10.1007/978-3-030-65380-4_3

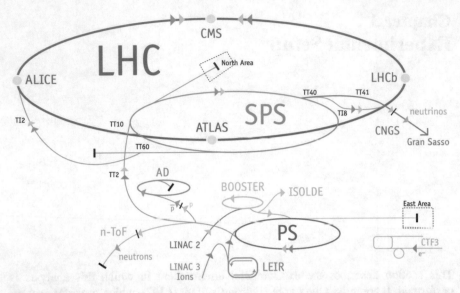

Fig. 3.1 The CERN accelerator complex and the four largest experiments at the interactions points of the LHC [6]

Fig. 3.1. Fueled by a hydrogen source, a duoplasmatron creates a proton beam with an energy of 90 keV [7]. Subsequently, the protons are accelerated in four stages starting with a linear accelerator (LINAC 2), followed by the Proton Synchrotron Booster (PSB), the Proton Synchrotron (PS), and the Super Proton Synchrotron (SPS) to energies of 50 MeV, 1.4 GeV, 25 GeV, and 450 GeV, respectively. The beam is eventually injected into the LHC after reaching the energy threshold of the SPS.

In the LHC, counter-rotating particles are accelerated by resonant waves of elec-tromagnetic fields, generated in eight radio-frequency (RF) cavities per beam. Each cavity operates in superconducting state at a temperature of 4.5 K to maintain an accelerating field of 2 MV per cavity at a frequency of $f_{RF} = 400.789$ MHz [5]. This results in a total, theoretical energy gain per revolution of 16 MeV per proton. In addi-tion, beam particles are confined to limited longitudinal intervals, so-called "buck-ets", due to the constant acceleration gating defined by f_{RF}. The maximum amount of possible buckets is expressed by the harmonic number $h = f_{RF}/f_{rev} \approx 35640$, with the revolution frequency f_{rev}. A bucket can either be empty or filled with a particle "bunch". The exact bunch configuration depends on the PS and SPS injection cycles and is altered to control interaction rates [2, 4].

Along the beam pipes, 1232 dipole magnets are installed to bend particle motions effectively into a circular trajectory. The NbTi magnets operate at a temperature of 1.9 K at which they reach their superconducting state to generate magnetic fields with a strength of up to 8.3 T. Further multipole magnets ensure that bunches remain focused despite the electromagnetic repulsion of the charged beam particles. At the interaction points of the LHC, the beams can be confined to a diameter in the order of $\sim 10\,\mu$m.

The center-of-mass energy $\sqrt{s}$ of the collision of two identical beam particles is given by

$$\sqrt{s} = 2 \cdot E_{\text{beam}}, \tag{3.1}$$

whereas for hadron colliders the energy available to the hard parton scattering process is smaller and subject to parton distribution functions (cf. Sect. 2.2.1). A second key characteristic is the instantaneous luminosity L, which summarizes all quantities related to the collider configuration to calculate the interaction rate of a process with cross section σ as

$$\frac{dN}{dt} = \sigma \cdot L \tag{3.2}$$

$$\text{with} \quad L = \frac{N_b^2 n_b f_{\text{rev}} \gamma_r F}{4\pi \epsilon_n \beta^*}, \tag{3.3}$$

with the number of particles per bunch N_b, the number of bunches per beam n_b, the revolution frequency f_{rev}, the relativistic gamma factor $\gamma_r = E/m$, the geometric reduction factor F due to the inclined beams profiles, the normalized transverse beam emittance ϵ_n, and the beta function at the interaction point β^* [5]. The design luminosity of the LHC is $10^{34}\,\text{cm}^{-2}\text{s}^{-1}$. The integrated luminosity $\mathcal{L}$ is a measure for the total amount of recorded data and transforms into the expected number of events for a given process with cross section σ through

$$N = \int \frac{dN}{dt}\, dt = \sigma \cdot \underbrace{\int L\, dt}_{:=\mathcal{L}} = \sigma \cdot \mathcal{L}. \tag{3.4}$$

Currently, seven experiments analyze the collisions delivered by the LHC. The CMS and ATLAS experiments are multipurpose detectors with focus on proton-proton collisions [8, 9]. Although their architecture and particle detection concepts are relatively different, they reach comparable performances and provide necessary, mutual cross checks for important physics measurements through independent results. Their extensive research programs include the illumination of the Higgs sector, high precision measurements of SM physics, and searches for signs of new physics such as supersymmetry, dark matter, or extra dimensions. Both experiments also study lead-proton and lead-lead collisions. The Large Hadron Collider beauty (LHCb) experiment is an longitudinally asymmetric detector equipped with a precise particle tracking system to analyze decays of long-lived b and c hadrons [10]. Especially studies of rare decays and CP violating processes enable sensitive tests of the SM. The Large Ion Collider Experiment (ALICE) investigates collisions involving heavy lead ions [11]. The resulting energy densities and temperatures are assumed to resemble the conditions shortly after the Big Bang, described by quasi-free quarks and gluons. One of the main purposes of the ALICE experiment is the detection of this

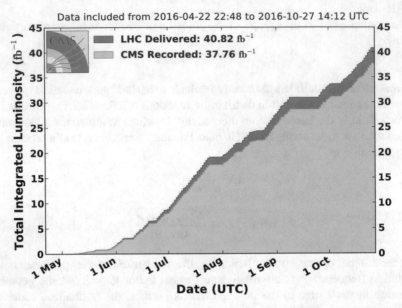

Fig. 3.2 Integrated luminosity over time provided by the LHC (blue) and recorded by the CMS experiment (yellow) [16]

so-called quark-gluon plasma and its evolution in the context of our understanding of strong interactions. The Total Elastic and Diffractive Cross Section Measurement (TOTEM) experiment studies charged particles originating from collisions inside the CMS detector in forward direction close to the beam line [12]. With detectors positioned on either side of the CMS experiment in distances of 145 and 220 m, its goal is the determination of the total proton-proton cross section and an independent measurement of the instantaneous luminosity. Neutral forward particles are analyzed by the Large Hadron Collider forward (LHCf) experiment, which consists of two detectors located on either side of the ATLAS experiment in a distance of 140 m to the interaction point [13]. It studies models of hadron interactions at very high energies, which, for example, are applicable in astroparticle physics experiments involving extensive air showers induced by cosmic rays. The Monopole and Exotics Detector at the LHC (MoEDAL) experiment is an extension of the LHCb experiment and searches for magnetic monopoles, i.e., yet unknown particles carrying magnetic charge [14].

In 2016, the LHC operated at a center-of-mass energy of $\sqrt{s} = 13$ TeV with 2040 bunches, 1.15×10^{11} particles per bunch, and a bunch spacing of 25 ns, which resulted in an instantaneous peak luminosity of $L = 0.77 \times 10^{34}$ cm^{-2}s^{-1} [15]. In total, an integrated luminosity of 40.82 fb^{-1} was provided at the interaction point in the center of the CMS experiment, of which 37.76 fb^{-1} was recorded [16]. The data provisioning and recording progress over time is depicted in Fig. 3.2.

3.2 The Compact Muon Solenoid

The CMS experiment is a large multipurpose detector for investigating proton-proton collisions provided by the LHC at high energies and instantaneous luminosities [8]. It also studies collisions involving lead atoms in a dedicated program. However, as it is not relevant for this thesis, it is not subject of the following discussions. The detector is located close to the city of Cessy, France, approximately 100 m under ground at the interaction point five of the LHC. Currently, around 5200 scientists, engineers, and students from 199 institutes in 45 countries take part in a huge collaborative effort, to push the frontiers of our understanding of nature. Except for neutrinos which interact only very weakly, the CMS detector is designed to measure all stable particles[1] produced in collisions events. In addition, the accuracy of energy and momentum measurements as well as the spatial resolution allow for the precise reconstruction of intermediate, unstable particles and provide access to the analysis of a broad range of physics processes.

To describe positions and directions of measured objects, the CMS experiment uses a coordinate system that is depicted in Fig. 3.3. Its origin is placed in the center of the detex ctor where the two proton beams collide. The x-axis points to the center of the LHC while the y-axis is oriented upwards. The two axes span the transverse plane and the azimuthal angle ϕ is defined as zero in direction of the x-axis. Consequently, the z-axis is perpendicular to the transverse plane parallel to the beam axis with positive, counter-clockwise direction. Instead of using the polar angle θ to denote directions, the pseudorapidity relative to the z-axis is defined as

$$\eta = -\ln\tan\left(\frac{\theta}{2}\right) = -\frac{1}{2}\ln\left(\frac{p + p_z}{p - p_z}\right) = \text{artanh}\frac{p_z}{p}, \tag{3.5}$$

where p is the absolute momentum of a particle and p_z is its signed z-component. The transverse component is denoted by p_T. Pseudorapidity differences are Lorentz-invariant and hence, they do not depend on longitudinal boosts of particular particles, which is important for hadron collision experiments as the interacting partons carry different momenta (cf. Sect. 2.2.1).

The CMS detector has a length of 21.6 m, a diameter of 14.6 m, and a mass of about 14.000 tons [8, 17]. Its structural design follows a cylindrical approach with multiple subdetectors aligned layerwise around the beam axis, symmetrically to the primary interaction point. With each component physicists, aim to measure different particle attributes to efficiently identify their type and kinematic properties. An illustration of the architecture is shown in Fig. 3.4. It comprises the following subdetector components starting from the center toward the outside.

- The inner tracking system consists of pixel detectors and silicon strip detectors to identify charged particles and to measure their trajectories, momenta, and charge

---

[1] Here, stable refers to relativistic particles that do not decay while traversing the detector.

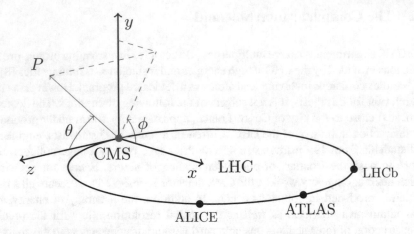

Fig. 3.3 The coordinate system used by the CMS experiment. The x-axis points to the center of the LHC, the y-axis is oriented upwards, and the z-axis is defined to be parallel to the beam line in counter-clockwise direction

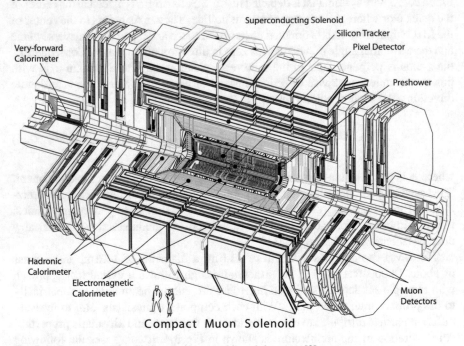

Fig. 3.4 Overview of the CMS experiment and its subdetectors [8]

signs. The two components are installed as close as possible to the interaction point to achieve the best possible spatial resolution.

- The calorimetry system comprises an electromagnetic calorimeter (ECAL), a preshower detector, a hadronic calorimeter (HCAL), and a very-forward calorimeter to measure energies and directions of traversing particles. The ECAL detects particles that interact electromagnetically, i.e., mostly electrons, positrons, and photons, whereas the HCAL is designed for the detection of charged and uncharged hadrons, i.e., protons, neutrons, pions, and kaons.

- A solenoid magnet is used to generate a magnetic field with a strength of up to 4 T to bend the trajectories of electrically charged particles, which enables the measurement of particle momenta in the tracker [8, 17]. With a length of 12.5 m and a diameter of 6.0 m, it is the largest solenoid magnet ever constructed. Its coil is made of superconducting NbTi and operates at a temperature of 4.7 K to maintain a resistance-free electrical current. To compress and return the magnetic flux outside the magnet into the detector volume, three layers of massive iron return yoke are installed which weigh about 10,000 tons in total.
- The muon system is installed between the layers of the iron return yoke. Its purpose is the identification of muons and the precise measurement of their momenta and charge signs.

The following sections discuss the above mentioned detector components in more detail and loosely follow Refs. [8, 17, 18]. Additionally, the concepts of the detector's data acquisition system are described in Sect. 3.2.4 and the measurement of the instantaneous and integrated luminosities is explained in Sect. 3.2.5. The algorithms utilized for the reconstruction of physics objects are discussed in the subsequent Sect. 3.3.

3.2.1 Tracking System

The tracking system of the CMS detector is installed in close proximity to the interaction point of the two proton beams and detects transit positions of charged particles. The detection principle is based on silicon semiconductors provided with a bias voltage to create a depletion zone in the active detector material. Therefore, transits of charged particles induce electron-hole pairs, which create measurable currents that are digitized using powerful readout electronics. These resulting "hits" can be grouped into tracks using advanced pattern recognition algorithms (cf. Sect. 3.3.1) to reconstruct a trajectory per particle indicating both its origin, or "vertex", and its direction of travel. Challenges arise especially from the readout frequency of 4×10^7 Hz resulting from the bunch crossing of 25 ns, detector calibration and alignment, computational complexity of track reconstruction, and radiation damage caused by the high particle flux.

The layout of the tracking system is tailored to the expected particle flux at various distances from the interaction point. At small radii, a silicon pixel detector with high granularity is installed to resolve hits of neighboring particles. Silicon microstrip detectors are used in outer regions with reduced flux. The overall used material, comprising also support structures, cables, and inner cooling constructions, is reduced to minimize its influence on traversing particles. Depending on the pseu-

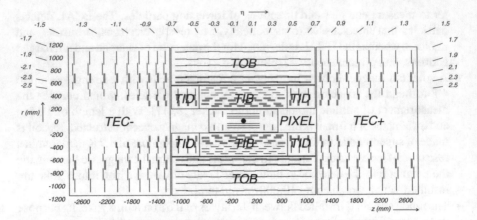

Fig. 3.5 Cross section of the tracking system layout of the CMS detector [8]. It shows the pixel detector in the center, followed by the inner barrel (TIB) and disk (TID) detectors, as well as the outer barrel (TOB) and endcap (TEC) subdetectors

dorapidity region, it is equivalent to 0.4 to 1.8 radiation lengths. In order to optimize radiation hardness, all tracking detector components are cooled down to a temperature of $-10\,°C$. In total, the CMS tracker covers a longitudinal region of $|\eta| < 2.5$, has a length of 540 cm, a diameter that extends to nearly 110 cm, and an active area of about $200\,m^2$, making it largest silicon tracking device ever constructed [17]. A structural view of the tracking system is shown in Fig. 3.5.

The inner pixel detector consists of three detector layers in the barrel region located at radii of 4.4, 7.3, and 10.2 cm, and two layers in the endcaps on each side of the detector at distances of ± 34.5 cm and ± 46.5 cm [17]. Each of the 66×10^6 pixels has a size of $100 \times 150\,\mu m^2$, which amounts to a net active area of $\sim 1\,m^2$. This granularity is required to resolve large numbers of traversing particles in high pileup conditions and to allow for a precise three-dimensional reconstruction of tracks and vertices. After the data taking period of 2016, the detector was replaced with four barrel and three endcap layers on each side, leading to a significant increase in track reconstruction efficiency and resolution [19].

The outer part of the tracking system consists of silicon microstrips which are sufficient to provide an unambiguous hit detection given the reduced particle flux and the resulting detector occupancy. The microstrip modules are organized in four regions with different layouts. The Tracker Inner Barrel (TIB) has a length of 130 cm and contains four layers of which the first two are made of "stereo" modules, consisting of successively arranged strips that are tilted against each other by an angle of 100 mrad. This setup, denoted by close pairs of lines in Fig. 3.5, results in a single-point resolution of 23 to 34 μm in transverse direction and of 23 μm in longitudinal direction. The Tracker Inner Disk (TID) covers the inner endcap regions and comprises three layers on each side of the detector with certain modules operating in stereo mode. The thickness of the strips in the TIB and TID is 320 μm with pitches between $80\,\mu m$ and $140\,\mu m$. This is surrounded by the Tracker Outer Barrel (TOB) in the central, and by the Tracker Endcaps (TEC) in the forward regions, which contain six and nine layers, respectively. Due to the flux reduction with larger radii, the strip thickness is increased to up to $500\,\mu m$. A total of 9.3×10^6 silicon microstrips provide an active detector area of approximately $\sim 198\,m^2$.

The efficiency of track reconstruction can be measured with muons originating from decays of Z bosons using a tag-and-probe method [20]. In 2016, the muon track efficiency was measured as $\sim 99\%$ in conditions with 30 pileup events, and slightly degrades to $\sim 96\%$ at 44 pileup events [21]. An alternative approach using charged pions originating from decays of $D^\pm$ mesons indicates a reasonable agreement between the track efficiencies measured in data and expected from simulation [21]. The spatial resolution of primary vertices was measured to be better than $14\,\mu m$ in transverse direction, and better than $19\,\mu m$ in longitudinal direction [22].

3.2.2 Calorimeters

The calorimeter system of the CMS detector is installed around the tracking system and is mostly enclosed by the solenoid. Its purpose is the precise energy measurement and full absorption of all particles except for muons and neutrinos. High granularity of its components is required for the accurate localization of energy deposits, which are combined with further subdetector information for the purpose of particle identification (cf. Sect. 3.3). It is divided into two subdetectors with different detection techniques and its layout is largely influenced by the dimensions of the solenoid it is surrounded by. The electromagnetic calorimeter (ECAL) measures photons and charged particles, mostly electrons and positrons, whereas the hadronic calorimeter (HCAL) is constructed for detecting strongly interacting particles such as protons, neutrons, pions, and kaons. In order to provide a reliable measurement of missing transverse energy, the calorimeter system is fully hermetic within a region of $|\eta| < 5$ with a maximized amount of used material to cover also the non-Gaussian tails of expected energy distributions [17]. The geometric calorimeter layout is depicted in Fig. 3.6.

The ECAL consists of 61.200 scintillating lead tungstate crystals (PbWO$_4$) in the cylindric barrel part (EB), and 7.324 crystals in each of the two endcaps (EE) [8, 23]. The EB has an inner radius of 129 cm and covers a region of $|\eta| < 1.479$, while the endcaps are placed in a distance of ± 314 cm to the interaction point and cover $1.479 < |\eta| < 3.0$. Lead tungstate has a short radiation length of 0.89 cm, which allows the crystals to be rather compact with a front cross section of $2.2 \times 2.2\,cm^2$ and a length of 23.0 cm in the EB, corresponding to ~ 26 radiation lengths. The EE crystals have slightly different dimensions with $2.86 \times 2.86\,cm^2$ front cross section and 22 cm length. The material is transparent for scintillation light, which is produced when traversed by charged particles. Its brightness provides a measure for the energy of the inducing particle. Due to the low light yield of lead tungstate, the light is amplified by avalanche photodiodes in the EB and vacuum phototriodes in the EE, respectively. In

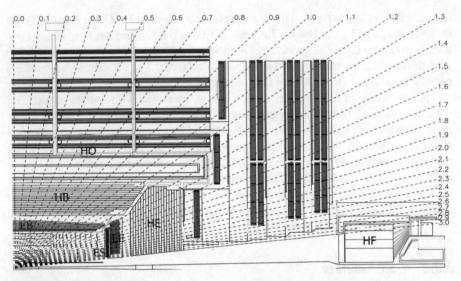

Fig. 3.6 Cross section of one quadrant of the CMS calorimeter system [8]. The electromagnet calorimeter (ECAL) consists of the ECAL Barrel (EB), the ECAL Endcaps (EE), and ECAL preshower (ES) components. The hadronic calorimeter (HCAL) comprises the HCAL Barrel (HB) and the HCAL Endcaps (HE), as well as the HCAL Outer (HO) detector beyond the solenoid and the HCAL Forward (HF) detector to measure objects close to the beamline

addition, preshower detectors (ES) are installed in front of the endcaps to cope with their reduced crystal granularity. They consist of two layers of silicon microstrip detectors followed by two lead absorber planes. This setup improves the spatial resolution of energy measurements in regions outside $|\eta| > 1.479$ and especially benefits the detection of neutral pions decaying into a pair of collimated photons, which can potentially be confused with a single photon. The energy resolution is estimated with test beam data by summing over the energy deposits of the 3×3 grid around a centrally hit crystal and can be expressed as [8]

$$\left(\frac{\sigma_E}{E}\right)^2 = \left(\frac{2.8\%}{\sqrt{E\,[\text{GeV}]}}\right)^2 + \left(\frac{12.0\%}{E\,[\text{GeV}]}\right)^2 + (0.3\%)^2. \tag{3.6}$$

It was reconfirmed by measuring electrons from Z boson decays in proton-proton collision data to 2% in the barrel, and to 2–5% in the endcap regions with an absolute energy scale precision of 0.4% and 0.8%, respectively [24].

The HCAL is a sampling calorimeter consisting of alternating layers of non-magnetic brass absorber and plastic scintillators [8, 17]. Its purpose is the absorption and energy measurement of strongly interacting particles that pass through the ECAL and thus, allowing for a basic division of particle types. Traversing particles create hadronic showers in the brass layers and induce detectable light in the subsequent scintillators, which is then guided by embedded wavelength-shifting fibers to the readout system. Owing to its limited size specified by the inner radius of the enclosing solenoid, the scintillator layers have a thickness of only 3.7 mm in order to maximize the amount of absorber material. The calorimeter is organized in four main components installed in different regions. The HCAL Barrel (HB) has a length of 9 m, an inner diameter of 6 m, and covers a range of $|\eta| < 1.4$. It is segmented in 2.304 towers with a size of $\Delta\eta \times \Delta\phi = 0.087 \times 0.087$. The HCAL Endcaps (HE) cover a range of $1.3 < |\eta| < 3.0$ and comprise 2.304 towers with sizes depending on the η region, ranging from 0.087 in $\Delta\eta$ and 5°–10° in $\Delta\phi$. To identify particles from tails of energetic hadron showers that pass through the HB and the magnet system, the HCAL Outer (HO) detector is installed outside the solenoid coil, covering a range of $|\eta| < 1.26$ and matching the ϕ segmentation of the muon system. It contains scintillators with a thickness of 10 mm and extends the effective absorption of the HB to approximately ten interaction lengths. Since the calorimeter is required to be as hermetic as possible, HCAL Forward (HF) subdetectors are placed on either side of the detector beyond the muon endcap systems. Unlike the other HCAL components, they are built from steel and quartz fibers causing narrower and shorter hadronic showers, which is well suited for the highly occupied forward regions. They have a distance of ± 11.2 m to the interaction point, a length of 165 cm, and cover a region of $3.0 < |\eta| < 5.0$. The energy resolution of the HCAL is designed to be approximately [25]

$$\left(\frac{\sigma_E}{E}\right)^2 = \left(\frac{100\%}{\sqrt{E\,[\text{GeV}]}}\right)^2 + (4.5\%)^2 \tag{3.7}$$

and was assessed and confirmed in combination with ECAL information to [26]

$$\left(\frac{\sigma_E}{E}\right)^2 = \left(\frac{85\%}{\sqrt{E\,[\text{GeV}]}}\right)^2 + (7.0\%)^2 . \tag{3.8}$$

Using proton-proton collision data recorded in 2016, the reconstructed energy response is found to be in excellent agreement between simulated and recorded data events [27]. Additionally, the impact of the magnetic field on the HCAL response is measured as 8–9% and is accounted for by energy calibrations [26].

3.2.3 Muon Detectors

The muon system constitutes the outermost component of the CMS detector and is interlaced with the iron return yoke of the magnet system [8, 17, 28]. Ideally, it is only traversed by muons and neutrinos, which, however, are not measurable. Therefore, hits detected by the muon system are combined with information of the tracking system to identify muons and measure their momenta and charges. The measurement principle is based on gaseous detectors, which produce electron avalanches through gas ionization when traversed by charged particles. The architecture of the muon

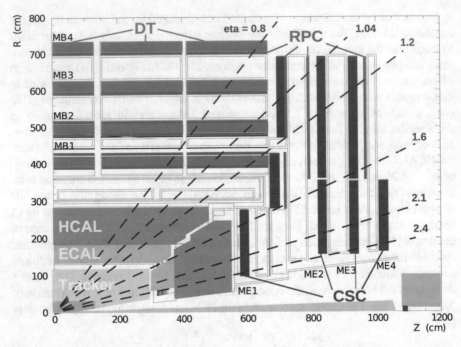

Fig. 3.7 Cross section of one quadrant of the CMS muon system [29]. The barrel region uses interlaced gaseous Drift Tube (DT) chambers and Resistive Plate Chambers (RPC), whereas the endcaps rely on a combination of Cathode Strip Chambers (CSC) and RPCs

system comprises three different detector approaches in two regions and is illustrated in Fig. 3.7.

The barrel region, $|\eta| < 1.2$, contains four layers of a total of 250 Drift Tube (DT) chambers complemented by five layers of Resistive Plate Chambers (RPC). The first three DT chambers contain twelve planes of aluminum drift tubes of which eight are responsible for measuring coordinates in the transverse $r-\phi$ plane and the remaining four planes function in the longitudinal $r-z$ plane. The outermost chambers only measure in $r-\phi$ direction with eight planes of drift tubes each. The endcaps are equipped with 486 Cathode Strip Chambers (CSC) up to a region of $|\eta| < 2.4$ as well as three RPC layers up to $|\eta| < 1.6$. The CSCs yield faster drift times as well as a three-dimensional spatial resolution since their cathode strips are segmented and positioned almost perpendicularly with respect to the anode wires. Furthermore, this design stabilizes the operation in the forward regions, which are usually congested by a high particle flux and subject to the strong, inhomogeneous magnetic field.

The single point resolution varies from 80–120 μm in the DT system, to 40–150 μm in the CSC, and to 0.8–1.2 cm in the RPC system [30, 31]. The DT and CSC systems offer a better spatial resolution of the muon tracks while the RPC detectors exhibit a faster response and deliver accurate timing information, which can be utilized for the purpose of triggering. Therefore, the approaches are employed in conjunction in the respective detector regions to achieve an optimal resolution for both time and position measurements. The hit and track segment reconstruction efficiency for traversing muons was found to be in the range of 95–98% [30, 31]. Also, the measured performance could be reproduced with simulations of the muon system, confirming the high efficiency and its effectiveness regarding the discrimination of backgrounds [30, 31].

3.2.4 Trigger and Data Acquisition

At the instantaneous design luminosity of $L = 10^{34}\,cm^{-2}s^{-1}$ with bunch crossings occurring every 25 ns, the LHC provides approximately 10^9 proton-proton collisions per second at the interaction point inside the CMS experiment [8, 17]. Considering all event information provided by the detector and a corresponding storage size in the order of $\sim$1 MB per event [18], this frequency would lead to a data output of $\sim$950 TB per second. However, the data acquisition system is only capable of processing events intended for permanent storage with a throughput of about $\sim$100 MB per second. Therefore, it is necessary to filter out the majority of events and to keep, or "trigger", only those who exhibit signatures of processes that are in accordance with the CMS physics program. This crucial task is the purpose of the trigger and data acquisition system, which consists of four components: detector electronics, the readout system, dedicated hardware trigger processors (Level 1 / L1 trigger), and a computing cluster running software-based High-Level Triggers (HLT) [32].

Recorded detector data can be held in internal buffers for up to 3.2 μs. During this time, signals from front-end electronics transit to the L1 trigger hardware, which calculates the decision to keep or dismiss an event in less than 1 μs [17] and sends back the result. Therefore, it only exploits information from the calorimeter and muons systems with reduced granularity to construct "trigger primitive" objects such as muons, electrons, photons, jets, and global sums of transverse and missing transverse energy. The maximum output rate of the L1 trigger amounts to $\sim$100 kHz.

Events passing the L1 trigger are transferred to the HLT for further filtering. Its trigger decision calculation is based on more sophisticated object constraints with increased precision including information from the inner tracking system. The HLT software is efficiently structured into so-called "paths" that aim to trigger events with various different topologies and physics objects but share common information to increase performance. An event is stored if it is accepted by at least one trigger path and discarded as soon as possible once it is established that no path will accept it. The output frequency of the HLT is in the order of $\sim$100 Hz. Depending on the instantaneous luminosity and pileup conditions, the rate of certain trigger paths may exceed the capabilities of the HLT hardware or the subsequent data storage system. In that case, the path is "pre-scaled" with a certain factor, e.g. 50%, meaning that it is only deployed for a fraction of events while the rest is a-priori rejected.

3.2.5 Luminosity Determination

The determination of luminosity is an essential ingredient for ensuring high data quality over a large quantity of events under constant measurements conditions. The measure of the online, real-time collision rate is used for monitoring beam and detector performance while the integrated luminosity over time, describing the absolute amount of recorded events for high-level physics analyses, is determined offline.

In the CMS experiment, five methods are employed to assess the luminosity based on event rates recorded with different detector components [8, 33]. The HF calorimeter, the Fast Beam Conditions Monitor (BCM1f), and the Pixel Luminosity Telescope (PLT) use fast readout electronics running asynchronously with respect to the main CMS readout system and give access to the real-time event rate and instantaneous luminosity. This is complemented by the silicon pixel detector and the DT chambers in the barrel of the muon system, which feature a low occupancy as well as a good stability over time, and function as part of the standard data acquisition system. Therefore, their recorded data can be evaluated and filtered offline by quality assurance measures, allowing for an accurate determination of the integrated luminosity.

The pixel detector is chosen as the primary offline luminometer for the 2016 data-taking period as it yields the best precision and a good pileup linearity up to 150 interactions per bunch crossing [33]. The measurement uses the Pixel Cluster Counting (PCC) method, which relates the instantaneous luminosity L within an event to the number of recorded pixel clusters $N_{clusters}$. The number of pixels is large enough to argue that a single pixel is not hit by two or more tracks of the same event, leading to a linear relation $L \propto N_{clusters}$ that can be parametrized as [33]

$$L = \frac{f_{rev}}{\sigma_{vis}^{PCC}} \times N_{clusters}, \tag{3.9}$$

with the LHC revolution frequency f_{rev} and the visible PCC cross section σ_{vis}^{PCC}. The method uses events recorded with an inclusive "zero-bias" trigger that only takes into account the expected bunch crossing rather than any specific event activity. The integrated luminosity is then inferred by summing over all zero-bias events. σ_{vis}^{PCC} is subject to a dedicated measurement using "van der Meer" (vdM) scans [34, 35].

The vdM scans performed with the CMS detector require a separate machine setup apart from the configuration deployed during data-taking [33]. During a scan, the relative transverse separation of the two beams is modified stepwise while monitoring the number of measured clusters in the pixel detector. This dependence is well described by a double-Gaussian curve parametrized by σ_{vis}^{PCC}, which can be extracted in a fit. The calibration performed by vdM scans might be affected by time, number of pileup interactions per bunch crossing, and the bunch filling scheme of the LHC. Therefore, scans are repeated to estimate the uncertainty on the absolute luminosity measurement, which is quoted to be 2.5% for data taken in 2016 [33].

3.3 Reconstruction of Physics Objects

The readouts of all channels of the subdetectors of the CMS experiment, as discussed in the previous Sect. 3.2, are transformed into binary hit information and calibrated energy deposits. In association with their three-dimensional positions in the detector coordinate system, these measurements are combined to form hypotheses of stable particles in terms of energy, momentum, flight trajectory, and, to some extent, particle type. This procedure is referred to as reconstruction of physics objects.

Processes of interest for the presented analysis contain electrons, muons, neutrinos, and multiple jets of which a subset is likely to originate from the decay of distinguishable, long-lived *b* hadrons. As neutrinos interact only very weakly, they are not detectable but rather impair the transverse energy balance. Therefore, the resulting amount and direction of missing transverse energy (MET) provides an important handle to assess neutrino activity in the event. This section explains how these physics objects are identified and reconstructed using an algorithm, the so-called Particle Flow (PF) algorithm, that takes measurements of all subdetector components into consideration.

3.3.1 The Particle Flow Algorithm

The purpose of the PF algorithm is the identification and reconstruction of individual particles by exploiting measurements of all available subdetectors [36]. The multitude of reconstructed particles in each proton-proton collision provides a full event description involving photons, electrons, muons, hadronically decaying tau leptons, jets, and derived quantities such as MET with unprecedented accuracy. In addition, the algorithm enables the reconstruction of the primary interaction vertex as well as the identification of particles from pileup events, allowing for efficient mitigation methods.

The PF algorithm can be structured into three consecutive steps. Firstly, information of subdetectors is gathered and preprocessed into so-called PF objects. They comprise tracks of charged particles within the inner tracking system, clusters of energy deposits measured in the calorimeters, and tracks of hits in the muon systems. Secondly, the findings in the three detector regions are spatially correlated with a link algorithm. Lastly, particle hypotheses are inferred and derived objects such as primary and secondary vertices, jets, and MET are computed. Important aspects of the three steps are described in the following paragraphs. An extensive reference of the PF algorithm is given in Ref. [36].

Track and Cluster Finding

Hits in the inner tracking system are reconstructed iteratively using a Combinatorial Track Finder (CTF) [20]. The CTF uses a Kalman filter (KF) [37] extended by pattern recognition and track fitting techniques [38, 39]. Starting from seeds of pairs or triplets of hits, track candidates are extrapolated to the outer layers of the tracking system and compatible hits are assigned using the quality of a fit to the expected flight path. The iterative approach commences by reconstructing relatively prominent tracks, e.g. from charged particles with large momentum. For subsequent iterations, hits related to these tracks are removed to decrease combinatorial complexity for less energetic or greatly displaced tracks. The computation time to reconstruct all tracks of a standalone $t\bar{t}$ event on a commercial CPU amounts to $\sim$0.7 s, and quickly increases with the number of pileup events to, e.g., $\sim$3.6 s at eight pileup events [20].

In order to exploit the full calorimeter granularity, the clustering of energy deposits is carried out separately in the ECAL barrel and endcaps, in the HCAL barrel and endcaps, and in the two preshower layers of the ECAL [36]. Cells exhibiting a local energy maximum over a certain seed threshold are considered clustering starting points and neighboring cells are iteratively added if they yield an energy deposit exceeding twice the noise level. In case of ambiguities in the cluster-cell association, energies are shared according to a distance measure. Finally, to distinguish particular energy deposits induced by multiple particles in these "topological clusters", the deposit profile is fit to a Gaussian-mixture model within an expectation-maximization algorithm [36].

Tracks in the muons system are obtained in two stages [28]. First, track segments are formed by clustering hits measured by the DT or CSC subdetectors (cf. Sect. 3.2.3). In the second step, these segments serve as seeds for pattern recognition techniques exploiting readouts of the full muon system, i.e., DT, CSC, and RPC components. The resulting object is called a "standalone-muon" track. More details on muon reconstruction are presented below (cf. Sect. 3.3.2).

Link Algorithm

The link algorithm connects tracks and clusters obtained in the previous step using distance measures. Owing to the quadratic computational complexity, only nearest neighbor pairs of elements as considered using a k-dimensional tree search [40]. As neutral particles can only be measured in calorimeters, the linking procedure is required to be as precise as possible to distinguish between calorimeter clusters from charged and uncharged particles.

Tracks measured in the inner tracker are linked to HCAL and ECAL barrel clusters in the (η, ϕ) plane, and to ECAL endcap and preshower clusters in the (x, y) plane. When several HCAL clusters are matched to the same track, or when several tracks are linked to the same ECAL cluster, only the combination with the smallest distance is kept. Otherwise, all combinations are retained. Links between HCAL and ECAL clusters are established in a similar manner. Ambiguity resolution is applied when multiple HCAL clusters are connected to the same ECAL cluster, or if several ECAL clusters are linked to the same preshower cluster. Finally, relations between inner tracks and standalone muon tracks are established, which is part of the muon reconstruction procedure (cf. Sect. 3.3.2).

Particle and Primary Vertex Reconstruction

The conclusive reconstruction of particles relies on PF objects and the geometrical links established between them. The procedure iteratively identifies physics objects in a well-defined order and removes associated PF objects for further steps.

First, muons with high evidence and quality are reconstructed and the expected energy deposited in calorimeter clusters is subtracted. Subsequently, electrons are identified using tracks that are matched to ECAL clusters. To account for the emission of bremsstrahlung photons, energies of ECAL clusters compatible with an extrapolated tangent of the track are added to the four-momentum of the electron hypothesis. Charged hadrons are built from links of the remaining tracks and HCAL clusters. In addition to the geometric links, ambiguities are resolved by comparing energy and momentum information. Track momenta that significantly exceed their connected calorimetric energy and that are also related to activity in the muon system, are identified as low quality muons. Finally, neutral hadrons and photons are inferred from calorimeter clusters that are associated to either no tracks, or to tracks whose momentum information is incompatible with the respective cluster energy. A higher energy in the HCAL is interpreted to stem from a neutral hadron, whereas an excess in the ECAL indicates a photon.

The location of primary interaction vertices can be exploited for the association to tracks and particles. After the identification of the "signal" vertex denoting the proton-proton collision of interest, all other vertices are attributed to pileup collisions and objects connected to them are discarded. The reconstruction of vertices consists of three steps. First, reconstructed tracks are selected with criteria regarding their number of silicon pixel and microstrip hits as well as their fit quality during the track finding procedure. Second, tracks are clustered using a deterministic annealing (DA) algorithm [41] based on the z-coordinate of their closest approach to the beam spot, resulting in an arbitrary number of collision vertices per bunch crossing. In the last step, the position of each vertex is determined using an adaptive fit method [20, 42]. The vertex that yields the highest sum of associated scalar transverse momenta is identified as the signal vertex, hereafter simple denoted as *the* primary vertex.

3.3.2 Muons

The PF algorithm reconstructs muons using information from both the inner tracking system and the muon detectors [28, 29, 36]. Two different muon reconstruction approaches are employed to provide a good description over a large momentum range.

The definition of "tracker muons" commences with inner tracks fulfilling $p > 2.5\,\mathrm{GeV}$ and $p_T > 0.5\,\mathrm{GeV}$ as measured by the central silicon tracking system, and extrapolating them to the outer muon system [36]. Multiple scattering processes with detector material are taken into account. If a track matches at least one muon segment, the object is identified as a tracker muon whose momentum corresponds to that of the inner track.

High energetic muons often penetrate more than one muon detector plane, causing hits in multiple segments. These hits are combined to form standalone muon tracks as discussed above (cf. Sect. 3.3.1). Opposed to tracker muons, "global muons" are built by extrapolating standalone muon tracks inwards to the central silicon tracker. A link to an inner track is established if it yields a high compatibility as determined by a Kalman filter approach. The momentum reconstruction is performed using both detector measurements in a fit to a global muon track, improving the momentum resolution with respect to a tracker-only fit [36].

Muons defined by the PF algorithm rely on both approaches, which are able to identify 99% of all muons produced within the acceptance of the CMS detector. If two candidates within one event share the same inner track, they are considered as a single muon to resolve ambiguities. Various selection criteria are applied to both collections to suppress fake background contributions while retaining a high identification efficiency and momentum resolution [31, 43]. An important quantity is the muon "isolation", which describes the activity related to other particles in a cone with radius $\Delta R = 0.3$ in the (η, ϕ) plane around the muon. By requiring that the sum of transverse momenta from tracks and energy deposits in the calorimeters should not exceed $\sim 10\%$ of the muon's transverse momentum, false identification of energetic hadrons can be adequately rejected.

The performance of the algorithms to identify and reconstruct muons as explained above was measured using $Z/\gamma^* \rightarrow \mu^+\mu^-$ events in collision data taken in 2016 [31, 43, 44]. For tight selection criteria, the identification efficiency is well above $\sim 95\%$ for $p_T > 20\,\text{GeV}$ with excellent agreement between data and simulation.

3.3.3 Electrons

The precise identification and reconstruction of electrons is essential for the calibration of detector responses as well as for the data-driven validation of various analysis methods. Compared to the muon reconstruction, however, two sources of potential inefficiencies emerge. While muons are measured in a dedicated subdetector, electrons are mainly identified using charged particle tracks and ECAL clusters, which are also subject to other types of particles. This circumstance results in a reduced electron identification efficiency on the one hand, and in a potentially false identification of photons or hadrons as electrons, so-called "fake" electrons, on the other hand. Secondly, the strong magnetic field causes charged particles to emit bremsstrahlung photons with a power proportional to $\propto (E/m)^4$. Interactions with tracking detector material lead to additional emissions of bremsstrahlung. Unlike for muons, the resulting energy loss is potentially significant and leads to kinked trajectories that affect track finding and linking to HCAL clusters [23, 36].

For the purpose of performance optimization, electrons are reconstructed using a combination of a standalone approach and the global PF algorithm [45]. Due to bremsstrahlung effects, the energy deposits of electrons in the ECAL are spread perpendicular to the magnetic field of the CMS detector. Basic clusters are built as explained above, starting from cells with large energies. They are combined to so-called "super-clusters" (SC) along the ϕ-direction and their position is obtained by an energy-weighted mean over all associated cells. Electron tracks are seeded with the results of the common track finding procedure using a combinatorial Kalman filter (KF) (cf. Sect. 3.3.1). However, their description is rather inaccurate in case of a significant amount of bremsstrahlung, causing non-Gaussian track deviations. Therefore, a Gaussian sum filter (GSF) is employed, which is able to follow sudden changes along the flight path. The two sets of seeds are combined and selected by a multivariate approach as electron seeds. While the track building is achieved with a combinatorial KF method, the track parameters are obtained by fitting the collected hits using a GSF fit. The energy loss in each layer is modeled using a mixture of Gaussian distributions. Finally, the electron momentum is determined by linking track and ECAL information depending on the spatial bremsstrahlung profile [36, 45].

The electron identification and tracking efficiencies are thoroughly monitored in every data-taking period. Using a tag-and-probe method in $Z/\gamma^* \to e^+e^-$ events recorded in 2016, the track finding efficiency is found to be in the order of ~95% over wide ranges of $25\,\text{GeV} < p_T < 500\,\text{GeV}$ and $|\eta| < 2.1$, with excellent agreement between real data and simulations [45, 46]. For tight selection criteria, the identification efficiency in data amounts to ~60% for $p_T > 20\,\text{GeV}$, and to ~80% for $p_T > 60\,\text{GeV}$.

3.3.4 Jets

Collimated jets of stable, reconstructed particles that are not identified as prompt photons, isolated leptons, or particles from pileup events give access to kinematic quantities of partons involved in the hard scattering process (cf. Sect. 2.2.1). In theoretical perspective, this relation holds as long as the jet clustering is independent of additional soft emissions and collinear particle splittings, referred to as infrared and collinear safety (IRC). Commonly used, IRC-safe jet clustering algorithms are the k_T [47], anti-k_T [48], and Cambridge-Aachen algorithms [49].

Jets used in the presented analysis are clustered with the anti-k_T algorithm and a distance parameter $\Delta R = \sqrt{\Delta\eta^2 + \Delta\phi^2} = 0.4$ [36, 48, 50]. The implementation used by the CMS experiment is based on the FASTJET package [51]. PF particles that appear to originate from interaction vertices other than the primary one are excluded by means of "charged hadron subtraction" (CHS) [52]. The anti-k_T clustering performs a sequential object merging where the order is defined through a distance measure

$$d_{ij} = \begin{cases} \min(p_{T,i}^{-2}, p_{T,j}^{-2}) \cdot \frac{\Delta\eta_{ij}^2 + \Delta\phi_{ij}^2}{R^2}, & i \neq j \\ p_{T,i}^{-2}, & i = j \end{cases} \tag{3.10}$$

between the pair of four-momenta of particles i and j, initialized with all particles considered for clustering. In case the minimal value d_{ij} results from two separate objects ($i \neq j$), their four-momenta are added and handled in subsequent iterations as a single object. Otherwise ($i = j$), the object leading to a minimal d_{ii} is excluded from further steps and considered a jet. Therefore, the anti-k_T algorithm first clusters hard and close particles before continuing with soft and distant ones, which yields desirable experimental and theoretical features such as robust reconstruction of the jet axis and a circular shape in the (η, ϕ) plane.

The jet energy scale and resolution in both real data and simulated events are subject to a manifold calibration procedure [50]. First, the energy offset induced by pileup events is corrected by comparing simulations of di-jet events with and without pileup modeling. Residual corrections between data and simulations are derived with the "random cone" (RC) method in events recorded with a zero-bias trigger. Subsequently, additional response corrections for various jet sizes are computed and applied to data and simulation. For data, residual corrections are determined via multijet events as a function of η, and with $Z/\gamma^* \rightarrow e^+e^-/\mu^+\mu^-$ +jets events, dependent on p_T. Finally, jet flavor corrections are derived from simulations by comparing predictions of complementary event generators, PYTHIA (v6.4) [53] and HERWIG++ [54]. Corrections to b-tagged jets are cross-checked with data in Z/γ^*+b events.

The jet identification efficiency was measured to be close to $\sim$100% using 2.3 fb^{-1} of collision data recorded in 2015 at $\sqrt{s} = 13$ TeV [55]. In addition, energy calibration uncertainties were estimated on the full dataset recorded in 2016 [56]. A breakdown of the uncertainties as functions of p_T and η of the jets is shown in Fig. 3.8. For central jets with $p_T > 30$ GeV, the total uncertainty is well below $\sim$2% and dominated by the jet flavor dependence with minor contributions from pileup offset, absolute, and relative scale corrections. Regarding the dependence on η, the total uncertainty is flat in the region $|\eta| < 2.5$ and determined as $\sim$2.5% for jets with $p_T \approx 30$ GeV. Again, the jet flavor description dominates in the central detector whereas pileup modeling and methodology-related uncertainties gain importance in more forward regions.

3.3.5 Missing Transverse Energy

Missing transverse energy (MET) measured in a proton-proton collision event is a key quantity to obtain information about particles that leave the experiment undetected such as neutrinos and, potentially, particles postulated by BSM theories. For instance, the presented analysis uses a MET threshold to select events with sufficient neutrino activity owing to the expected $t\bar{t}H$ signal process signature and to suppress backgrounds from QCD multijet processes. Under the assumption that the transverse momenta of the partons involved in the hard scattering process are negligible, the momenta of reconstructed final state particles are balanced in the transverse plane [36]. Therefore, missing transverse energy $\not{E}_T$ is defined as the magnitude of the

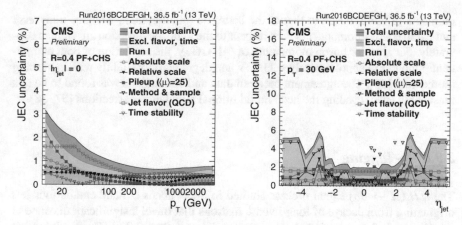

Fig. 3.8 Jet energy calibration uncertainties in the data-taking period 2016 for central jets as a function of p_T (left) and for jets with $p_T \approx 30\,\text{GeV}$ as a function of η (right) [56]

negative vectorial sum of all reconstructed particle momenta, $\vec{p}_T^{\,\text{miss}}$, as

$$\vec{p}_T^{\,\text{miss}} = -\sum_i^{\text{particles}} \vec{p}_{T,i} \tag{3.11}$$

$$\not{E}_T = |\vec{p}_T^{\,\text{miss}}|. \tag{3.12}$$

Experimentally, a hermetic calorimetry system and an excellent muon reconstruction efficiency are essential preconditions. The measurement of the longitudinal component is not possible as initial momentum fractions of interacting partons are a-priori unknown (cf. Sect. 2.2.1).

The determination of MET is sensitive to the reconstruction efficiencies and accuracies of all other particles, and therefore relies on the performance of the PF algorithm. Especially mis-identified or mis-reconstructed muons with large transverse momenta affect the quality of the MET description to a great extent [36]. These events may occur when muons induced by cosmic rays traverse the CMS detector in coincidence with a bunch crossing. They can be rejected by requiring the closest proximity of the muon trajectory to the beam axis to be within 1 cm, while differentiating displaced muon tracks from leptonic b hadron decays [36]. Other sources are the incorrect association of muon detector hits to inner tracks, leading to large differences between available estimates of the muon momentum, and "fake" muons caused by punch-through of highly energetic hadrons beyond the magnet system.

The performance of MET reconstruction was estimated using recorded collision data [57, 58]. Scale and resolution were studied in $Z/\gamma^* \rightarrow e^+e^-/\mu^+\mu^- +$jets events with good Z boson reconstruction accuracy but no intrinsic source of missing transverse momentum. Hence, the MET quality can be assessed by comparing momenta of the Z boson to that of the hadronic recoil system, whereas transverse and parallel MET components with respect to the Z boson direction are studied separately. For a vector boson momentum of $\sim$100 GeV, the resolution of both components amounts to approximately $\sim$23 GeV and degrades only slightly with increasing momentum [58]. The agreement between data and simulations was found to be at a reasonable level, evading the need for additional simulation corrections [57, 58].

3.3.6   b-*Tagging*

The $t\bar{t}H$ ($H \rightarrow b\bar{b}$) signal process studied in the analysis at hand entails four jets originating from decays of long-lived b hadrons that travel a significant distance in the detector before decaying at a secondary vertex (cf. Sect. 2.2.4). The displacement with respect to the position of the primary vertex, typically $\sim$1 cm at a momentum of $\sim$100 GeV, can be resolved by exploiting the high resolution of the inner tracking system. This approach gives rise to the identification of b-flavored jets, called "b-tagging", and constitutes a powerful tool in many high-energy physics analyses. For example, by distinguishing b-flavored jets from those induced by other quark flavors or gluons, the combinatorial complexity in the reconstruction of events with high jet activity is significantly reduced. Also, background contributions from processes involving non-b-flavored jets can be suppressed.

Secondary vertices are reconstructed in an Inclusive Vertex Finding (IVF) [59] procedure, which functions similar to the reconstruction of primary vertices. Selected tracks are required to have a maximum (η, ϕ) distance of 0.3 relative to the jet axis, a longitudinal distance below 0.3 cm, and a transverse momentum above 0.8 GeV. After a clustering of the tracks, initial positions of secondary vertices are obtained in an adaptive fit. Ambiguities in the track association to either the primary or a secondary vertex are resolved by evaluating a compatibility measure based on the angle between the track and the flight direction with respect to the vertex. Finally, a secondary vertex is retained if it is associated to at least two tracks, and its position is fit conclusively.

Various b-tagging methods are deployed in the CMS experiment. Following a synchronization across multiple working groups participating in the $t\bar{t}H$ effort, the presented analysis utilizes the Combined Secondary Vertex algorithm in its second revision (CSVv2) [59, 60]. It combines information of displaced tracks and secondary vertices associated to a jet within a multivariate technique implemented as a supervised multilayer perceptron with a single hidden layer [59]. In total, the discrimination power of 22 variables is exploited and aggregated into a single discriminator output. A jet is attributed a b tag if the discriminator output exceeds a threshold value. This value, the so-called "working point", is usually defined by choosing a particular false positive rate of the tagger, i.e., the probability to erroneously identify a jet to have b-flavor.

The performance of the CSVv2 b-tagging algorithm was measured in simulations and compared to collision data recorded in 2016 [59]. Figure 3.9 shows the misidentification probability as a function of the b-tagging efficiency in simulated $t\bar{t}$

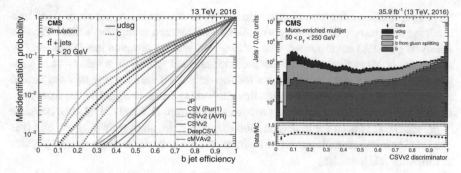

Fig. 3.9 Left: Misidentification probability versus b-tagging efficiency for various tagging algorithms as measured in simulated $t\bar{t}$ events [59]. At a misidentification probability of 1%, the CSVv2 algorithm has a tagging efficiency of ~64% against jets from gluons and u, d, and s quarks (blue solid line). Right: Comparison of the discriminator output of the CSVv2 algorithm between data and simulation in muon-enriched QCD multijet events [59]

events (left) and a comparison of the CSVv2 discriminator distribution as measured in recorded and simulated jets (right). At a misidentification probability of 1%, the tagging efficiency amounts to ~64% against jets from gluons and u, d, and s quarks, so-called "$udsg$ jets". Against c-flavored jets, the efficiency drops to ~33% owing to similarities regarding life time and decay characteristics of c hadrons. The comparison to jets in recorded collisions events yields a reasonable agreement in a large transverse momentum range $50\,\text{GeV} < p_T < 250\,\text{GeV}$. Residual differences are to be addressed by corrections that depend on the phase space of the particular analysis (cf. Sect. 6.4).

3.4 Software and Libraries

Physics data analyses investigating high-energy particle collisions, such as presented in this thesis, comprise the study of a vast quantity of recorded and simulated events to examine rare processes with sufficient statistical significance. In addition, sophisticated algorithms are essential for the fulfillment of a multitude of tasks, ranging from physics object reconstruction, to signal extraction procedures with multivariate techniques, as well as management of particular analysis workflows and the dependencies between them. To cope with the challenges posed by both scale and complexity in an effective and resource-efficient way, state-of-the-art physics data analyses rely on a considerable amount of software libraries.

This section summarizes the principal software tools that are employed to perform the presented analysis. First, the software and the event data model developed and used by the CMS experiment are briefly described. Then, the Physics eXtension Library (PXL) is introduced, which is used to represent and store collision events in a hierarchical structure with mother-daughter relations between particles. Furthermore, it allows for the encapsulation of parametrized analysis algorithms in a modular structure, which can be visualized and steered using the VISual Physics Analysis (VISPA) environment. The subsequent section presents the TENSORFLOW library for constructing and evaluating computational graphs on a variety of heterogeneous hardware systems. It allows for the definition and training of deep neural networks, which are employed within this analysis to separate signal events from contributions of multiple background processes (cf. Chap. 7). Lastly, the LUIGI package is presented. It is used for managing and executing the analysis workflow, and it is extended to integrate with remote infrastructures available to the high-energy physics community (cf. Sect. 5.1).

3.4.1 CMS Software Framework

The software framework of the CMS experiment, CMSSW, serves a broad spectrum of applications. It is deployed in the high-level trigger of the detector for data-taking purposes, i.e., for the implementation of trigger paths, for detector alignment calibration tasks, for the production of simulated collision events, and in physics object reconstruction routines [18]. It also constitutes the foundation of subsequent high-level data analysis, which requires the framework to exhibit a high degree of flexibility.

The control and data flow of CMSSW is event-centric and based on the Event Data Model (EDM) [18]. A CMSSW program consists of configurable modules that are written in C++ and encapsulate particular analysis algorithms. Modules can be parametrized to encourage modular programming and code reusability. During execution, data is passed between them on an event-by-event basis and per event, a module can either reject it from further processing, analyze existing event content, or append new content to be analyzed in downstream modules. All information is stored in so-called "collections", which can describe a variety of event-related quantities such as hits, tracks, and energy deposits, but also high-level data like trigger decisions and reconstructed particle hypotheses. Dependencies between modules are defined by requirements to named collections of upstream modules. The resulting directed graph is inferred automatically and, if available, can be executed in a multi-threaded process with some modules operating simultaneously while others run sequentially. The configuration and steering of these graphs is enabled through the prototyping-friendly Python programming language. Furthermore, the full encapsulation of parameterizable analysis algorithms allows for the storage of detailed provenance information [61]. Internally, the data format is based on tree structures as provided by the ROOT toolkit [62].

The analysis at hand uses CMSSW in version CMSSW_8_0_26_patch2.

3.4.2 PXL and VISPA

The Physics eXtension Library (PXL) provides an experiment-independent collection of classes designed for applications in high-energy physics data analysis [63, 64]. Its key components comprise the means to organize event-related quantities in a hierarchical, extensible structure, and tools for constructing analysis sequences in a modular fashion following an event-by-event data flow. Using PXL, data can be read from and saved to disk through a dedicated, compressed I/O format. The so-called "pxlio" files are event-based and headless, and allow for the fast merging and splitting of files with standard UNIX utilities. While written in C++, all objects are also accessible through Python-bindings generated by the SWIG interface generator [65].

PXL provides interfaces for generic physics objects, e.g. `pxl::Particle` and `pxl::Vertex` classes with a variety of convenience methods, as well as container structures to describe and process them in logical, nestable groups, such as the `pxl::Event` and `pxl::EventView` classes. The concept of event views allows for a consistent description of different aspects of the same event. Common use cases are the separation of generator- and reconstruction-related information of simulated events, or the bookkeeping and retaining of varying reconstruction hypotheses of the identical event. In addition, objects located in the same container can be associated among each other via mother-daughter relations. They support various applications ranging from the modeling of particle interactions within event generator information similar to Feynman diagrams, to the representation of results of complex association and clustering algorithms. Furthermore, PXL allows objects to be attributed with additional information during run-time using so-called `pxl::UserRecords`. A user record is a key-value pair that maps a string identifier to a `pxl::Variant` object, which is able to hold data of various types. They are defined per object without introducing artificial constraints between them, resulting in a versatile and dynamic data structure.

The second key functionality of PXL is a module system for defining analysis sequences. A sequence contains multiple modules that are connected through a common interface by passing, for example, `pxl::Event` instances from one to another in a directed flow. For modeling arbitrary, branched control patterns, modules can define one or more input (`pxl::Sink`) and output (`pxl::Source`) ports. Data is passed along connections between two ports and can be handled appropriately in the module implementation. Moreover, algorithms encapsulated in modules are parameterizable, encouraging full code modularity and reusability. Besides C++ modules for performance-intensive algorithms, PXL also supports Python modules within the same sequence for fast prototyping and testing purposes. After a development cycle, analyses are executable using the `pxlrun` command-line tool, which allows for further configuration of module parameters at execution-time.

Nested physics data structures and modular analysis sequences as supported by PXL can be visualized with the VISual Physics Analysis (VISPA) software [63]. VISPA provides a graphical development environment for interactively designing,

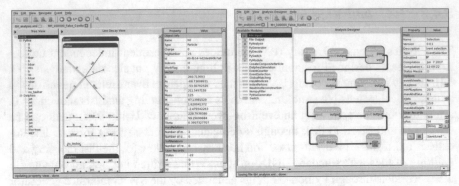

Fig. 3.10 Screenshots of the physics data browser (left) and graphical analysis designer (right) applications in VISPA

executing, and validating physics analyses, as demonstrated exemplarily in Fig. 3.10. The left-hand figure shows the data browser, which visualizes the content of PXL files on the basis of events. Particles are contained in event views and decay trees are visualized according to the relations between them. By clicking on a particular object, all its properties such as four-momentum information and user records can be inspected interactively. The right-hand figure depicts the visual analysis designer. Starting from the set of all available predefined and individually developed modules, an analysis sequence is built by selecting and connecting modules according to the respective task to be performed. Module parameters are configurable through a structured user interface. Once the analysis is saved in a human-readable format, execution and inspection of results can be performed right within the application.

The VISPA project provides an implementation of the analysis software that is accessible through a common web browser to take advantage of the numerous benefits inherent to web applications [66]. Following a typical server-client approach based on modern web technologies, analysis-specific software can be installed on a central server, which mediates and dispatches incoming requests to dedicated computing resources. The user interface provides the same functionality as described above and is accessible independent of device and location. An instance of the VISPA web service is deployed on a computing cluster hosted by the "III. Physics Institute A" at the RWTH Aachen University for research, outreach, and education [67, 68]. The resources attached to the setup comprise a dedicated GPU computing cluster, which was utilized for the training of deep neural networks employed in this analysis (cf. Chap. 7).

3.4.3 TensorFlow

TENSORFLOW is a software library for high performance numeric computation and symbolic differentiation [69]. Initially designed by Google, it was released in 2015 to the public and is developed in an open-source community effort ever since. Owing to its flexible structure, support for a multitude of platforms, and comprehensive math library, it is well equipped for applications in the area of machine learning, especially in the context of deep neural networks. It is interfaced to the commonly used array implementation provided by the NumPy package [70].

Using TENSORFLOW, a computational task is represented by a directed graph with nodes and edges denoting "operations" and "tensors", respectively. The example graph shown in Fig. 3.11 (left) describes the computation

$$C = \underbrace{\underbrace{W \cdot x}_{\text{"MatMul"}} + b,}_{\text{"Add"}} \tag{3.13}$$

where W, b, and x are input tensors with suitable shapes, and C is the output tensor. The purple boxes represent operations such as addition and matrix multiplication, whereas the arrows denote the direction of the graph, i.e., the *flow* of tensors. In addition to basic arithmetic operations, TENSORFLOW provides a variety of operations that are specially tailored for use in machine learning applications. Therefore, if applicable, operations are defined along with their symbolic derivation rules which influence the learning behavior of a network by means of the backpropagation algorithm (cf. Sect. 5.2.2).

The computation of a graph is initiated when the numerical result of one or more operations is requested, whereby TENSORFLOW automatically determines the subgraph to process. Here, three types of tensors are distinguished, namely "placeholders", "constants", and "variables". Placeholders usually mark the entry point for external information, such as input features for neural network evaluation, and must be provided manually. Variables and constants are initialized with a-priori values, whereas only variables can be subject to updates during graph processing and usually describe the trainable parameters of a machine learning model. Applied to the example introduced above, C is a function of x, given the variable tensors $\theta = (W, b)$, such that $C = C(x|\theta)$.

Similar to this rather elementary graph, large machine learning models often consist of compositions of vector and matrix operations whose computational performance can be optimized substantially using vectorization techniques. Modern graphics processing units (GPUs) exhibit highly parallel architectures, often equipped with thousands of processing cores that operate far more efficient for vectorized algorithms. Therefore, TENSORFLOW provides a "device placement" mechanism to distribute the actual computation of a virtual graph across physical devices, e.g. CPUs and GPUs, which is illustrated in Fig. 3.11 (right). After placing operations on particular devices, their execution is scheduled depending on the graph structure, and

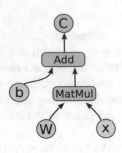
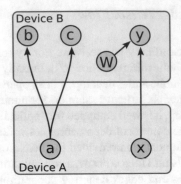

Fig. 3.11 Left: Example of a directed computational graph in TENSORFLOW [69] (image slightly altered). Nodes are represented by operations, whereas edges are denoted by tensors. Right: Placement of operations on different devices for performance optimization purposes, requiring the exchange of tensor data between them [59]

tensor states are transferred between them if necessary. Internally, GPU instruction sets are accessed through the CUDA parallel computing platform [71].

3.4.4 Luigi

LUIGI is a Python software package that provides a scalable design pattern for structuring large and complex workflows of arbitrary workloads [72]. Initially developed at Spotify, it became a community-driven, open-source project and is successfully deployed in both commercial and scientific applications. To include remote resources available in the context of high-energy physics research, LUIGI was extended along the development of this analysis. The concepts and capabilities of this extension are discussed in detail in Sect. 5.1. The following paragraphs introduce the fundamental building blocks of the LUIGI package.

Using LUIGI, an arbitrary workload, i.e., the elementary unit in an overarching workflow, is described as a "task". The purpose of a task is to produce a customizable set of outputs, denoted by so-called "targets". While targets usually represent local or remote files, they can, in principle, describe any type of stateful resource (e.g. a database entry). The sole core functionality of a target is to check and report its own existence. Therefore, a task is considered "complete", when all its output targets exist. Moreover, to alter the default behavior of a workload, tasks can expose and implement mandatory and optional "parameters". Significant task parameters, i.e., parameters that influence the content of output targets, should be encoded in the respective target locations, e.g. via distinctively integrating them into file paths. The resulting injective mapping is advisable in order to prevent ambiguities during checks of a task's completeness condition. Finally, tasks can "require" one or multiple other tasks to define a coherent workflow, where outputs of required tasks become

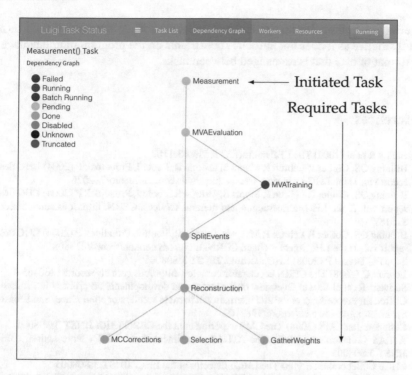

Fig. 3.12 Screenshot of the LUIGI web visualizer application, showing a hypothetical analysis workflow consisting of an initiated task and its recursive requirements in a directed acyclic graph (DAG). Edges visualize direct dependencies and node colors denote task statuses

"inputs" to the current task. As tasks can implement requirements depending on values of particular parameters, workflows can behave highly dynamically and adapt to a large variety of use cases.

A workflow that consists of interdependent tasks can be described as a directed acyclic graph (DAG), and visualized upon execution using a web application provided by LUIGI. An example is shown in Fig. 3.12 for a hypothetical analysis. The execution of a workflow is initiated by running a particular task. LUIGI recursively evaluates the completeness condition of all required tasks to infer the shape of the underlying DAG. Subsequently, tasks are scheduled according to their position in the DAG, whereby the number of parallel processed tasks as well as additional priority rules are configurable. As a result, LUIGI's stateful execution system is deterministic and resource-effective since it only processes what is really necessary.

For physics applications, workflow management tools such as LUIGI can give rise to considerable advantages. Large-scale analyses with complex dependency structures can be modeled via task workflows based on a design pattern consisting of only a few building blocks, while keeping overhead on a reasonably low level. Besides the clear separation of workloads and a consistent parameterization approach, all dependencies are programmatically expressed as part of the analysis with direct benefits for error prevention, documentation, and collaboration. At the same time, LUIGI is non-restrictive as it does not introduce constraints on the programming language or the format of data that is exchanged between tasks.

References

1. Bailey R et al (2002) The LEP collider. C R Phys 3:1107
2. Brüning OS, Collier P, Lebrun P, Myers S, Ostojic R, Poole J, Proudlock P (2004) LHC design report vol. 1: the LHC main ring. CERN. http://cds.cern.ch/record/782076
3. Brüning OS, Collier P, Lebrun P, Myers S, Ostojic R, Poole J, Proudlock P (2004) LHC design report vol. 2: the LHC infrastructure and general services. CERN. http://cds.cern.ch/record/815187
4. Brüning OS, Collier P, Lebrun P, Myers S, Ostojic R, Poole J, Proudlock P (2004) LHC design report vol. 1: the LHC injector chain. CERN. http://cds.cern.ch/record/823808
5. Evans L, Bryant P (2008) LHC machine. JINST 3:S08001
6. Lefèvre C (2008) The CERN accelerator complex. http://cds.cern.ch/record/1260465
7. Scrivens R, et al (2011) Overview of the status and developments on primary ion sources at CERN. In: Proceedings of the 2nd international particle accelerator conference, San Sebastian, Spain. http://cds.cern.ch/record/1382102
8. Collaboration CMS (2008) The CMS experiment at the CERN LHC. JINST 3:S08004
9. ATLAS Collaboration (2008) The ATLAS experiment at the CERN large hadron collider. JINST 3:S08003
10. LHCb Collaboration (2008) The LHCb detector at the LHC. JINST 3:S08005
11. ALICE Collaboration (2008) The ALICE experiment at the CERN LHC. JINST 3:S08002
12. TOTEM Collaboration (2004) Total cross-section, elastic scattering and diffraction dissociation at the large hadron collider at CERN: TOTEM technical design report. CERN. CERN–LHCC–2004–002. http://cds.cern.ch/record/704349
13. LHCf Collaboration (2006) LHCf experiment: technical design report. CERN, CERN–LHCC–2006–004. http://cds.cern.ch/record/926196
14. MoEDAL Collaboration (2009) Technical design report of the MoEDAL experiment. CERN, CERN–LHCC–2009–006. http://cds.cern.ch/record/1181486
15. Metral E (2016) LHC: status, prospects and future challenges. In: Proceedings of the 4th annual large hadron collider physics conference, Lund, Sweden. http://cds.cern.ch/record/2265090
16. CMS Collaboration (2016) CMS public luminosity results. https://twiki.cern.ch/twiki/bin/view/CMSPublic/LumiPublicResults
17. CMS Collaboration (2006) CMS physics: technical design report volume 1: detector performance and software. CERN-LHCC-2006-001. http://cds.cern.ch/record/922757
18. CMS Collaboration (2005) CMS computing: technical design report. CERN-LHCC-2005-023. http://cds.cern.ch/record/838359
19. CMS Collaboration (2012) CMS technical design report for the pixel detector upgrade. CERN-LHCC-2012-016. http://cds.cern.ch/record/1481838
20. CMS Collaboration (2014) Description and performance of track and primary-vertex reconstruction with the CMS tracker. JINST 9:P10009. arXiv:1405.6569 [physics.ins-det]
21. CMS Collaboration (2016) CMS tracking POG performance plots for year 2016. https://twiki.cern.ch/twiki/bin/view/CMSPublic/TrackingPOGPlots2016
22. CMS Collaboration (2016) Primary vertex resolution in 2016. CMS Performance Note CMS-DP-2016-041. http://cds.cern.ch/record/2202968
23. CMS Collaboration (1997) The CMS electromagnetic calorimeter project: technical design report. CERN-LHCC-97-033. http://cds.cern.ch/record/349375
24. CMS Collaboration (2013) Energy calibration and resolution of the CMS electromagnetic calorimeter in pp collisions at $\sqrt{s} = 7$ TeV. JINST 8:P09009. arXiv:1306.2016 [hep-ex]

25. CMS Collaboration (2010) Single-particle response in the CMS calorimeters. CMS Physics Analysis Summary CMS-PAS-JME-10-008. http://cds.cern.ch/record/1279141
26. CMS Collaboration (2010) Performance of the CMS hadron calorimeter with cosmic ray muons and LHC beam data. JINST 5:T03012. arXiv:0911.4991 [physics.ins-det]
27. CMS Collaboration (2016) HCAL energy reconstruction performance. CMS Performance Note CMS-DP-2016-071. http://cds.cern.ch/record/2235509
28. CMS Collaboration (1997) The CMS muon project: technical design report. CERN-LHCC-97-032. http://cds.cern.ch/record/343814
29. CMS Collaboration (2012) Performance of CMS muon reconstruction in pp collision events at $\sqrt{s} = 7$ TeV. JINST 7:P10002. arXiv:1206.4071 [physics.ins-det]
30. CMS Collaboration (2013) The performance of the CMS muon detector in proton-proton collisions at $\sqrt{s} = 7$ TeV at the LHC. JINST 8:P11002. arXiv:1306.6905 [physics.ins-det]
31. CMS Collaboration (2018) Performance of the CMS muon detector and muon reconstruction with proton-proton collisions at $\sqrt{s} = 13$ TeV. JINST 13:P06015. arXiv:1804.04528 [physics.ins-det]
32. CMS Collaboration (2017) The CMS trigger system. JINST 12:P01020. arXiv:1609.02366 [physics.ins-det]
33. CMS Collaboration (2017) CMS luminosity measurements for the 2016 data taking period. CMS Physics Analysis Summary CMS-PAS-LUM-17-001. http://cds.cern.ch/record/2257069
34. van der Meer S (1968) Calibration of the effective beam height in the ISR. CERN-ISR-PO-68-31
35. Grafström P, Kozanecki W (2015) Luminosity determination at proton colliders. Prog Nucl Phys 81:97
36. CMS Collaboration (2017) Particle-flow reconstruction and global event description with the CMS detector. JINST 12:P10003. arXiv:1706.04965 [physics.ins-det]
37. Frühwirth R (1987) Application of Kalman filtering to track and vertex fitting. Nucl Instrum Meth A 262:444
38. Billoir S (1989) Progressive track recognition with a Kalman like fitting procedure. Comput Phys Commun 57:390
39. Billoir S, Qian S (1990) Simultaneous pattern recognition and track fitting by the Kalman filtering method. Nucl Instrum Meth A 294:219
40. Bentley JL (1975) Multidimensional binary search trees used for associative searching. Commun ACM 18:509
41. Rose K (1998) Deterministic annealing for clustering, compression, classification, regression and related optimisation problems. Proc IEEE 86:2210
42. Frühwirth R, Waltenberger W, Vanlaer P (2007) Adaptive vertex fitting. J Phys G 34:N343
43. CMS Collaboration (2017) Muon identification and isolation efficiency on full 2016 dataset. CMS Performance Note CMS-DP-2017-007. http://cds.cern.ch/record/2257968
44. CMS Collaboration (2016) Performance of muon reconstruction including alignment position errors for 2016 collision data. CMS Performance Note CMS-DP-2016-067. http://cds.cern.ch/record/2229697
45. CMS Collaboration (2015) Performance of electron reconstruction and selection with the CMS detector in proton-proton collisions at $\sqrt{s} = 8$ TeV. JINST 10:P06005. arXiv:1502.02701 [physics.ins-det]
46. CMS Collaboration (2014) Electron and photon performance in CMS with the full 2016 data sample. CMS Performance Note CMS-DP-2017-004. http://cds.cern.ch/record/2255497
47. Catani S, Dokshitzer YL, Seymour MH, Webber BR (1993) Longitudinally invariant K_t clustering algorithms for hadron hadron collisions. Nucl Phys B 406:187
48. Cacciari M, Salam GP, Soyez G (2008) The anti-k_t jet clustering algorithm. JHEP 04:063. arXiv:0802.1189 [hep-ph]
49. CMS Collaboration (2009) A Cambridge-Aachen (C-A) based jet algorithm for boosted top-jet tagging. CMS Physics Analysis Summary CMS-PAS-JME-09-001. http://cds.cern.ch/record/1194489

50. CMS Collaboration (2017) Jet energy scale and resolution in the CMS experiment in pp collisions at 8 TeV. JINST 12:P02014. arXiv:1607.03663 [hep-ex]
51. Cacciari M, Salam GP, Soyez G (1896) FastJet user manual. Eur Phys J C 72:1896. arXiv:1111.6097 [hep-ph]
52. CMS Collaboration (2014) Pileup removal algorithms. CMS Physics Analysis Summary CMS-PAS-JME-14-001. http://cds.cern.ch/record/1751454
53. Sjöstrand T, Mrenna S, Skands P (2006) PYTHIA 6.4 physics and manual. JHEP 05:026. arXiv:hep-ph/0603175
54. Bahr M, et al (2008) Herwig++ physics and manual. Eur Phys J C 58:639. arXiv:0803.0883 [hep-ph]
55. CMS Collaboration (2016) Jet algorithms performance in 13 TeV data. CMS Physics Analysis Summary CMS-PAS-JME-16-003. http://cds.cern.ch/record/2256875
56. CMS Collaboration (2018) Jet energy scale and resolution performance with 13 TeV data collected by CMS in 2016. CMS Performance Note CMS-DP-2018-028. http://cds.cern.ch/record/2622157
57. CMS Collaboration (2017) Missing transverse energy performance in high pileup data collected by CMS in 2016. CMS Performance Note CMS-DP-2017-028. http://cds.cern.ch/record/2275227
58. CMS Collaboration (2019) Performance of missing transverse momentum reconstruction in proton-proton collisions at $\sqrt{s} = 13$ TeV using the CMS detector. JINST 14:P07004. arXiv:1903.06078 [hep-ex]
59. CMS Collaboration (2018) Identification of heavy-flavour jets with the CMS detector in pp collisions at 13 TeV. JINST 13:P05011. arXiv:1712.07158 [physics.ins-det]
60. CMS Collaboration (2013) Identification of b-quark jets with the CMS experiment. JINST 8:P04013. arXiv:1211.4462 [hep-ex]
61. Jones CD, et al. (CMS Collaboration) (2010) File level provenance tracking in CMS. J Phys: Conf Ser 219:032011
62. Brun R, Rademakers F (1997) ROOT - an object oriented data analysis framework. Nucl Instrum Meth A 389:81
63. Erdmann M, et al (2012) A development environment for visual physics analysis. JINST 7:T08005. arXiv:1205.4912 [physics.data-an]
64. Erdmann M, et al, Physics eXtension library. https://vispa.physik.rwth-aachen.de/pxl
65. Beazley DM (1996) SWIG: an easy to use tool for integrating scripting languages with C and C++. In: Proceedings of the 4th annual USENIX Tcl/Tk workshop, Monterey, USA, vol 4, p 15
66. Erdmann M, Rieger M, et al (2017) Experiment software and projects on the web with VISPA. J Phys: Conf Ser 898:072045. arXiv:1706.00954 [physics.data-an]
67. Erdmann M, et al (2014) A field study of data analysis exercises in a bachelor physics course using the internet platform VISPA, vol 35, p 035018. arXiv:1402.2836 [physics.ed-ph]
68. Erdmann M, et al (2018) The VISPA internet-platform in deep learning applications. J Phys: Conf Ser 1085:042044
69. Abadi M, et al (2016) TensorFlow: large-scale machine learning on heterogeneous distributed systems. Prelim White Pap (2016). arXiv:1603.04467 [cs.DC]
70. Oliphant TE (2006) A guide to NumPy. Trelgol Publishing, ISBN: 9781517300074
71. Nickolls J, Buck I, Garland M, Skadron K (2008) Scalable parallel programming with CUDA. Queue-GPU Comput 6:40
72. The Luigi Authors. Luigi Documentation. https://luigi.readthedocs.io

Chapter 4
Analysis Strategy

This section presents an overview of the strategy that is applied in this analysis of proton-proton collisions recorded by the CMS experiment in 2016 to search for $t\bar{t}H$ ($H \to b\bar{b}$) production. Several challenges have to be addressed in order to measure the rare signal process with considerable sensitivity in the presence of dominating backgrounds from $t\bar{t}$ contributions. While these challenges stem from physical considerations, they directly affect the pursued measurement strategy and the technical design of the analysis. The following paragraphs describe the emerging challenges as well as the key concepts employed for their accomplishment.

The subsequent sections are organized as follows. First, the physical conditions entailed by the search for $t\bar{t}H$ ($H \to b\bar{b}$) production are summarized with focus on the interplay with irreducible $t\bar{t}$+hf backgrounds. The measurement strategy is presented thereafter, which is composed of the event selection strategy and resulting physics processes to consider, the event classification and categorization approach, and the concluding signal extraction method. The section concludes with an outline of technical considerations that follow from the design of the measurement strategy.

4.1 Physical Conditions

This analysis searches for the production of $t\bar{t}H$ events with Higgs boson decays into a pair of bottom quarks, and single-lepton (SL) or dilepton (DL) decays of the $t\bar{t}$ system. The nominal final state consists of six jets and an electron or muon in the single-lepton channel, or four jets and two oppositely charged electrons or muons in the dilepton channel. Both channels contain four jets originating from decays of b hadrons as well as a significant amount of missing transverse energy due to undetectable neutrinos.

The dominant background results from the production of top quark pairs with additional heavy-flavor jets in the final state ($t\bar{t}$+hf). Especially events with two

© The Author(s), under exclusive license to Springer Nature Switzerland AG 2021
M. Rieger, *Search for tt̄H Production in the H → bb̄ Decay Channel*, Springer Theses,
https://doi.org/10.1007/978-3-030-65380-4_4

additional bottom quarks, $t\bar{t}+b\bar{b}$, constitute an irreducible[1] background contribution. However, due to detector acceptance, jet reconstruction and selection criteria, as well as misidentification efficiencies in b-tagging algorithms (cf. Sect. 3.3.6), further background processes such as $t\bar{t}+b/\bar{b}$, $t\bar{t}+2b$, $t\bar{t}+c\bar{c}$, $t\bar{t}+$lf, or single $t/\bar{t}$ production are to be considered. A detailed discussion of the $t\bar{t}H$ ($H \rightarrow b\bar{b}$) event and all relevant background processes is presented in Sect. 2.2.4. The statements on background characteristics made in the following paragraphs mainly target $t\bar{t}+$hf processes and in particular the $t\bar{t}+b\bar{b}$ process. However, although to a lesser extent, they also apply for the remaining backgrounds.

Due to their similar final state content, $t\bar{t}H$ ($H \rightarrow b\bar{b}$) and $t\bar{t}+$hf events cannot be fully separated from each other with selection cuts in the phase space of available observables. This contamination plays a central role in the design of the analysis strategy. The resulting physical measurement conditions can be divided into five, partially interdependent aspects.

1. **Rare signal process** With a cross section of $\sigma_{t\bar{t}H,b\bar{b},\mathrm{SL+DL}} = 98.4$ fb [1], the signal process is $\sim$2800 times less likely than inclusive $t\bar{t}$ production with $\sigma_{t\bar{t},\mathrm{SL+DL}} = 276.9$ pb [2–14] at a center-of-mass energy of $\sqrt{s} = 13$ TeV. Although only $\sim$15% of these $t\bar{t}$ events are produced with additional heavy-flavor jets (cf. Sect. 6.2), a vast amount of events with similar final state and kinematic topology remains. This disparity must be taken into account in the design of the event selection procedure.

2. **Irreducible backgrounds** As outlined above, contributions from $t\bar{t}+$hf processes constitute an important source of background events in this analysis. While especially $t\bar{t}+b\bar{b}$ processes are irreducible with respect to the nominal final-state content, effects due to detector acceptance, fiducial cuts, and reconstruction algorithms require the consideration of further processes in the overall background modeling.

3. **Uncertain backgrounds** As discussed in Sect. 2.2.4, the definition of $t\bar{t}+$hf background processes is based on the number of b and c hadrons found in jets clustered on generator level after parton showering and hadronization processes are simulated via phenomenological methods. Currently, their expected rates cannot be constrained by either higher-order theoretical calculations nor experimental analyses with accuracies better than 35% [15–18]. Corresponding rate uncertainties on theses processes exceed the nominal signal yield by approximately one order of magnitude (cf. Sect. 6.2). Conservative 50% normalization uncertainties are considered in this analysis, uncorrelated for all $t\bar{t}+$hf contributions.

4. **Complex final state** With six jets and one lepton in the single-lepton, and four jets and two leptons in the dilepton channel, as well as missing transverse energy in both cases, the final state is rather complex. The reconstruction of the Higgs boson from two jets entails combinatorial ambiguities, leading to 15 (6) possible hypotheses in the single-lepton (dilepton) channel, which increase rapidly with additional jets. Owing to the non-detectable neutrinos, a full event reconstruction

---

[1] Irreducible in terms of the final state content.

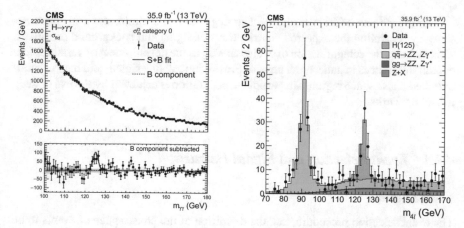

Fig. 4.1 Visible and narrow peaks in mass distributions of the $H \to \gamma\gamma$ analysis (left) [21], and of the $H \to ZZ^* \to 4l$ analysis (right) [22]. Due to low (b) jet mass resolution, a comparable structure is not expected to be observable in the $m_{b\bar{b}}$ spectrum of the $t\bar{t}H$ ($H \to b\bar{b}$) analysis

requires sophisticated methods whose processing is particularly demanding in the dilepton channel [19].

5. **Low mass resolution**   Unlike the Higgs boson decay channels $H \to \gamma\gamma$ and $H \to ZZ^* \to 4l$, the decay into a pair of bottom quarks, $H \to b\bar{b}$, exhibits a relatively low mass resolution [20]. Figure 4.1 exemplarily demonstrates the prominent peaks found in the mass spectra of the aforementioned channels [21, 22]. Due to the low dijet mass resolution, a comparable structure is not expected to be observed in the presented analysis, and therefore, the extraction of results must be subject to statistical inference methods (cf. Sect. 5.3).

These physics-induced conditions have a direct impact on the measurement strategy of this analysis. The general approach including the concepts of event selection, process classification, and signal extraction is presented in the following.

4.2 Measurement Strategy

The search for $t\bar{t}H$ ($H \to b\bar{b}$) production in this analysis is designed along a well-defined measurement strategy that can be divided into two parts. They are discussed in the following.

The first part describes the general measurement procedure consisting of the essential, necessary steps to perform the analysis and obtain results. It commences with the selection of events, followed by the application of residual corrections to simulations, and eventually leads to the signal extraction technique, which is based on a parametrized fit to data.

The second part of the measurement strategy aims to optimize the analysis sensitivity by separating the expected contributions of signal and background processes. This includes the categorization of events as well as the development of variables that exploit differences in underlying physics processes to define signal- and background-enriched phase space regions. In particular, separation is achieved by the use of deep neural networks.

4.2.1   Event Selection and Signal Extraction

Event Selection

The event selection procedure, i.e., the definition of the phase space of events to be analyzed, serves two purposes. On the one hand, it is employed to exclude or suppress background processes while retaining as many signal events as possible. For searches, the ratio between expected signal and background events S/B, or sometimes $S/\sqrt{B}$, is a good estimator to assess the selection quality. On the other hand, selection criteria are to be defined such that phase space regions are excluded in which the simulation is found to describe recorded data with insufficient precision. Moreover, selection cuts on physics objects are synchronized with recommendations from Physics Object Groups (POG) and Physics Analysis Groups (PAG) within the CMS experiment. They provide centrally measured identification and reconstruction efficiencies, as well as estimates on the overall agreement between data and simulation of related quantities.

Section 6.3 describes the full set of selection criteria in detail. As a result, five groups of background processes are relevant for this analysis: top quark pair production with additional jets ("$t\bar{t}+X$"), single top quark production ("single $t/\bar{t}$"), vector boson production with additional jets ("V+jets"), the production of two vector bosons ("diboson"), and top quark pair production in association with a vector boson ("$t\bar{t}+V$") (cf. Sect. 2.2.4). Contributions from QCD multijet background were found to be negligible in previous analyses that studied a similar phase space after incorporating selection criteria on the amount of missing transverse energy [23–28].

Corrections

Subsequent to the event selection, a set of corrections is applied to simulated events in order to mitigate residual differences with respect to data. These corrections either change individual event and object attributes, or they only account for deviations observed for a collection of events, usually addressed with a reweighting of events based on binned distributions of certain quantities. The latter approach introduces the concept of event weights which, by construction, affect the distributions of all involved event variables. The correction procedure follows guidelines proposed by CMS POGs and PAGs, and is discussed in Sect. 6.4.

Eventually, the quality of the background modeling can be determined by comparing various distributions of simulated and recorded collision events in control regions of the analysis. These regions are excluded in the actual measurement procedure, while being sufficiently similar to infer meaningful conclusions. At this stage, recorded data is not evaluated in the analysis region to ensure an unbiased progress. This approach is referred to as "blinding".

Signal Extraction

As explained in Sect. 4.1, an excess in the invariant mass spectrum of two b-tagged jets is not expected to be observable. Therefore, the signal extraction procedure relies on a fit of a simulated, parametrized variable distribution to data. Since the distribution is binned, the underlying fit model is constructed according to an ensemble of counting experiments, where the expected event yield in each bin is described by a Poisson distribution. The model parameters, which are all determined in the fit, contain the parameter of interest, i.e., the $t\bar{t}H$ signal strength modifier $\mu = \sigma/\sigma_{SM}$, as well as a set of "nuisance" parameters that mediate the effect of systematic uncertainties. Physics knowledge on systematic uncertainties is included via appropriate prior probability distributions, whereas no assumptions are made on μ; its prior distribution is flat with no preconditioned range. The fit method is described in detail in Sect. 5.3.

In the context of a small amount of expected signal events compared to all backgrounds, this construction entails a leverage effect, which implies a crucial dependence of μ on the nuisance parameters. First, any residual modeling discrepancy between data and simulation is compensated during the fit primarily by adjusting central nuisance parameter values. Careful monitoring of these nuisance "pulls" is therefore required as they, in turn, strongly affect the measured value of μ. Second, the determination of each nuisance parameter is intrinsically subject to a measurement uncertainty. A parameter is said to be "constrained" if the uncertainty, i.e., the width of its posterior probability distribution, is found to be smaller than assumed a-priori. Consequently, large posterior uncertainties of nuisance parameters cause undesired, high variances on the measurement of μ. Therefore, the ability to measure and constrain particular nuisance parameters is directly connected to the sensitivity of the analysis.

The signal extraction procedure is first performed for the blinded analysis. Observed data is replaced by the "Asimov" dataset, i.e., an artificial dataset in which all observable quantities are approximated from simulation [29]. Contributions of $t\bar{t}H$ signal events are included by setting $\mu = 1$. Posterior expectation values of all fit model parameters are identical to their assumed a-priori values. Constraints on nuisance parameter distributions can be interpreted to assess the consistency of the statistical model and the convergence behavior of the fit. After thorough validation, the analysis is unblinded and the fit is repeated with actual data.

Three quantities are defined to describe the expected and observed results of the analysis. For expected values, the statistical fit model distinguishes between the "background-only hypothesis", i.e., the absence of signal by requiring the central signal strength to be $\mu = 0$, and the "signal hypothesis" assuming a central signal strength modifier of $\mu = 1$.

Upper limit   For the exclusion of hypotheses, upper limits on μ at 95% confidence
level are constructed using the CL_s technique [30]. The expected upper limit typically refers to the background-only hypothesis. A value greater than one signifies
that, even in the presence of signal in Asimov data, the expected background is
capable of explaining the data on its own. An analysis is considered sensitive, if an
expected limit below one is obtained to disfavor the background-only hypothesis
with at least 95% confidence.

Best fit value   The best estimate of μ is obtained after full convergence of the fitting
procedure. Its central value and uncertainty can be transformed to the measured
cross section of the signal process. For the blinded analysis with Asimov data,
the signal hypothesis is utilized in order to derive an estimate for the expected
accuracy of the measurement of μ.

Significance   The significance describes the excess of data over the expected background contribution and is expressed in Gaussian standard deviations (σ) via conversion of p-values. The terminology of searches for new physics introduces two
thresholds at 3σ and 5σ above which an analysis reveals the "evidence" or the
"observation" of a process, respectively.

4.2.2   Event Categorization and Process Discrimination

The general measurement strategy, as discussed in the previous section, is self-
contained as it comprises the necessary steps to obtain results. In general, the fit
of the underlying statistical model to data could be performed on a binned distribution of an arbitrary event variable. However, as motivated above, the sensitivity
of background-dominated analyses strongly depends on the ability to constrain systematic uncertainties through nuisance parameters in the fit. This is achieved by
separating signal and background yields across different bins of the fitted distribution.

Event Categorization

The first step is the categorization of events based on properties related to the expected
final-state content. Its purpose is the grouping of events with similar characteristics into logically and topologically distinct categories that, by construction, exhibit
different signal and background process compositions. A common categorization
scheme utilizes the number of measured jets and b tags [25–28]. While the jet identification efficiency is close to ~100% (cf. Sect. 3.3.4), the b-tagging efficiency depends
on the applied tagging algorithm, its working point, and the analysis phase space.
In events with more than two expected b jets $\hat{n}_b$, category transition probabilities
become relevant. Given a tagging algorithm with an identification efficiency ε and a
misidentification efficiency $\bar{\varepsilon}$, the probability to measure n_b b tags in an event with
a fixed number of jets, $\hat{n}_b$ b jets, and $\hat{n}_l$ non-b jets, is

Table 4.1 Probabilities to measure n_b b tags in events with six jets and a true number of b jets $\hat{n}_b$. Values are subject to rounding and follow Eq. 4.1 with tagging and mis-tagging efficiencies of $\epsilon = 64\%$ and $\bar{\epsilon} = 1\%$, respectively. Diagonal values mark the probabilities to measure the correct number of b tags

Events with 6 jets	True b jets, $\hat{n}_b$			
	2	3	4	5
Measured b tags, n_b — 2	41.1%	43.7%	31.5%	19.0%
3	1.6%	26.7%	37.6%	33.8%
4	< 0.1%	0.8%	17.2%	30.2%
5	< 0.1%	< 0.1%	0.3%	10.9%

$$p(n_b, \hat{n}_b, \hat{n}_l \mid \epsilon, \bar{\epsilon}) = \sum_{i=a}^{b} \binom{\hat{n}_b}{i} \epsilon^{\hat{n}_b - i} (1 - \epsilon)^i \cdot \binom{\hat{n}_l}{i - \Delta n_b} \bar{\epsilon}^{i - \Delta n_b} (1 - \bar{\epsilon})^{\hat{n}_l + \Delta n_b - i}$$

$$(4.1)$$

$$\text{with} \quad \Delta n_b = \hat{n}_b - n_b \tag{4.2}$$

$$a = \max(0, \ \Delta n_b) \tag{4.3}$$

$$b = \min(\hat{n}_b, \ \hat{n}_l + \Delta n_b). \tag{4.4}$$

For $\epsilon = 64\%$ and $\bar{\epsilon} = 1\%$ (cf. Sect. 3.3.6), the probabilities to measure certain numbers of b tags in events with six jets is exemplarily shown in Table 4.1. The diagonal values represent the combined probabilities to assign an event correctly to the appropriate category. This probability amounts to only 17.2% for single-lepton $t\bar{t}H$ ($H \rightarrow b\bar{b}$) events with 4 b jets. As a result, and contrary to the underlying conception, the construction of categories using the number of measured b tags does not create topologically distinct groups that can be exploited to achieve separation between signal and background processes. This analysis pursues a refined categorization scheme which is explained below.

Discriminating Variables

The second step to separate signal and background contributions relies on more complex features of the involved processes, which are often extracted by means of multivariate analysis methods (MVA) [31]. MVA methods combine multiple variables of reconstructed events and exploit correlations among them to identify differences in the provided phase space densities. Subsequently, these differences are projected onto a single dimension to construct one or more variables that discriminate between signal and background processes. These discriminating variables are then used in the fit as part of the signal extraction procedure.

Previous searches for $t\bar{t}H$ ($H \to b\bar{b}$) production [23–28] employed Boosted Decision Trees (BDT) [32] and the Matrix Element Method (MEM) [33–35]. While the former is a machine learning algorithm that requires a dedicated training process, the latter approach is physics-motivated and relies on the probability for a reconstructed event to stem from a hypothetical, partonic matrix element. Both methods perform a binary classification by distinguishing two classes of events, i.e., signal and background. The different physics processes that contribute to the overall background are treated equally.

In the statistical model of this analysis (cf. Sect. 8.1), the normalization uncertainties of the $t\bar{t}$+hf backgrounds are modeled conservatively by independent nuisance parameters per process. Since a binary classification leads to bins that are enriched with all backgrounds at once, these nuisance parameters interfere with each other during the fit and might impair its ability to infer constraints. As discussed above, this circumstance would effectively translate into a reduction of analysis sensitivity. Therefore, a different event discrimination and classification approach is introduced in the presented analysis.

Multi-class Classification with Deep Neural Networks

As discussed in the previous paragraphs, this analysis utilizes a newly developed event categorization scheme as well as a set of discriminating variables that are specially tailored for the search of $t\bar{t}H$ ($H \to b\bar{b}$) production. In fact, both approaches are results of the same algorithm, which is based on deep neural networks (DNNs). While the technological concepts of DNNs are explained in Sect. 5.2, a dedicated discussion of the event classification method is presented in Chap. 7. The following paragraphs summarize its significance for the measurement strategy.

The key mechanism of the categorization and process classification algorithm is a DNN that performs a multi-class classification. An illustration is shown in Fig. 4.2. Per event, a predefined set of input variables is fed into the network whose architecture and weights are subject to a training and optimization process. The network evaluates the variables and outputs a vector of six elements with a sum of one. Each value describes the probability for a certain event to originate from either the $t\bar{t}H$ signal, or one of the five $t\bar{t}$+X background processes. The distinct categorization into one of the six "classes" is derived according to the class with the highest output value, i.e., the most probable process. Simultaneously, the distributions of the six output units constitute powerful discriminating variables.

In each of the created process categories, a different discriminating variable distribution is selected for the signal extraction procedure. The configuration is as follows. For events that are categorized as $t\bar{t}H$, the variable corresponding to the $t\bar{t}H$ output value is chosen. The same choice is applied for the $t\bar{t}$+X categories and their corresponding outputs. The resulting mapping is unambiguous and avoids double-counting of events. By construction, the lower boundary of a variable in its respective category is $1/6 \approx 0.167$.

This architecture not only separates signal and background events, but also divides the latter into the individual contributions from different processes. As a consequence, dedicated bins are created that are mainly enriched with one of the relevant physics

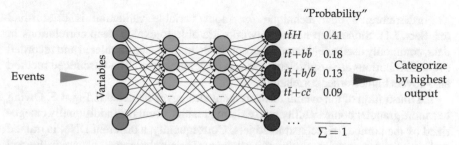

Fig. 4.2 Concept of the event categorization approach using DNNs that perform a multi-class classification. Each event is attributed six values that express its probability to originate from either the $t\bar{t}H$ signal, or one of the five $t\bar{t}+X$ background processes. An event is categorized by the process class that received the highest probability ($t\bar{t}H$ in the depicted example)

processes. Interference between the involved nuisance parameters is mitigated in the fitting procedure and stronger model constraints can be deduced. In turn, the analysis sensitivity is expected to improve.

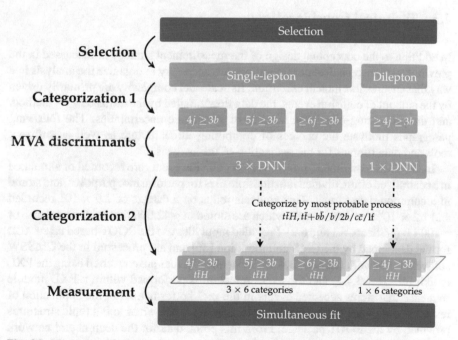

Fig. 4.3 Illustration of the categorization and measurement strategy developed for this analysis. After a selection in single-lepton and dilepton channels with a minimum number of b tags, events are initially categorized by the number of jets. In each category, Deep Neural Network (DNN) discriminants are trained to categorize events further by their most probable process. Finally, the measurement is performed through a fit of the expected DNN output distributions to data simultaneously in all 24 categories

Furthermore, a new technique for input variable validation is established (cf. Sect. 7.1). Since deep neural networks are able to exploit deep correlations in data, commonly used one-dimensional comparisons between simulated and recorded event distributions are insufficient to assess their agreement. The introduced method also covers higher-order correlations.

An illustration of the overall measurement strategy is depicted in Fig. 4.3. Owing to a more granular comparability with existing results, events are additionally categorized by the number of reconstructed jets. Consequently, a different DNN is trained and optimized in each jet multiplicity category. The single-lepton channel is divided into events with four $(4j)$, five $(5j)$, or more than six jets $(\geq 6j)$, whereas the dilepton channel requires four or more jets $(\geq 4j)$. Both channels treat electrons and muons equally and require at least three b tags. Thus, the combined b-tagging efficiency to accept an event with three (four) true b-flavored jets amounts to 27.5% (55.2%) (Eq. 4.1). Events with two b tags constitute control regions. In total, this results in 24 categories with corresponding discriminants that are employed simultaneously in the signal extraction procedure.

4.3 Technical Considerations

In addition to the conceptual design of the measurement strategy as discussed in the previous section, technical considerations are necessary to organize the analysis in a way that ensures its efficient execution. Its scale and complexity are primarily driven by the amount of collision events, the data flow entailed by the utilized MVA method, and the processing of a large number of systematic uncertainties. The following paragraphs motivate the choices of computing infrastructure as well as software tools and data formats for the realization of the analysis.

In the CMS experiment, proton-proton collision events are recorded or simulated in a centralized effort, divided into multiple files for performance purposes, and stored in a common data format. This analysis relies on a dataset of 2.9×10^9 recorded and 1.2×10^9 simulated events which are stored in $\sim$42.000 files with a total size of $\sim$100 TB (cf. Sects. 6.1 and 6.2). Provided input files use the ROOT-based miniAOD format developed by the CMS collaboration [36] and are processed in the CMSSW software framework (cf. Sect. 3.4.1). Subsequently, files are converted using the PXL toolkit (cf. Sect. 3.4.2) and the event selection is performed within a PXL module sequence that stores accepted events in the pxlio format. After the application of residual corrections to simulated events, the data is converted to flat tuple structures provided by the ROOT package. From this point, data for the deep neural network training is derived, process classification and event categorization are conducted, and histograms of important variable distributions are created for both plotting purposes and the final signal extraction procedure.

The majority of these workloads is processed on the Worldwide LHC Computing Grid (WLCG) [37], whereas DNN trainings are executed on a dedicated GPU computing cluster deployed by the "III. Physics Institute A" at the RWTH Aachen University [38].

The underlying data flow is designed to be as parallel as possible and is primarily dictated by three factors. First, the size of files is reduced after event selection and conversion among the aforementioned data formats. To optimize job run-times, file transfers, and CPU consumption, files are consecutively concatenated between particular analysis steps to achieve optimal file sizes for processing on WLCG infrastructure. Second, the DNN trainings are performed on 50% of simulated, nominal $t\bar{t}H$ and $t\bar{t}+X$ events, and are hence required to be merged into single files. After successful training, the DNNs are evaluated on the remaining simulated and all data events for subsequent analysis with high parallelism. Finally, this search is subject to a total of 80 systematic uncertainties (cf. Sect. 8.1), which cause the repetition of certain parts of the analysis with varied configurations.

Therefore, a novel workflow management system was created to cope with the scale and complexity of this analysis. It is based on the open-source software package LUIGI developed by Spotify (cf. Sect. 3.4.4) and establishes a generic analysis design pattern with integrated support for interchangeable remote resources. Details of the workflow software are explained in Sect. 5.1.

References

1. LHC Higgs Cross Section Working Group (2017) Handbook of lhc higgs cross sections: 4. deciphering the nature of the higgs sector. CERN Yellow Reports: Monographs. arXiv:1610.07922 [hep-ph], http://cds.cern.ch/record/2227475
2. Beneke M et al (2012) Hadronic top-quark pair production with NNLL threshold resummation. Nucl Phys B 855:695 arXiv:1109.1536 [hep-ph]
3. Cacciari M et al (2012) Top-pair production at hadron colliders with next-to-next-to-leading logarithmic soft-gluon resummation. Phys Lett B 710:612 arXiv:1111.5869 [hep-ph]
4. Bärnreuther P, Czakon M, Mitov A (2012) Percent level precision physics at the tevatron: first genuine NNLO QCD corrections to $q\bar{q} \rightarrow t\bar{t}$. Phys Rev Lett 109:132001 arXiv:1204.5201 [hep-ph]
5. Czakon M, Mitov A (2012) NNLO corrections to top-pair production at hadron colliders: the all-fermionic scattering channels. JHEP 1212:054 arXiv:1207.0236 [hep-ph]
6. Czakon M, Mitov A (2013) NNLO corrections to top pair production at hadron colliders: the quark-gluon reaction. JHEP 1301:080 arXiv:1210.6832 [hep-ph]
7. Czakon M, Fiedler P, Mitov A (2013) The total top quark pair production cross-section at hadron colliders through $O(\alpha_s^4)$. Phys Rev Lett 110:252004 arXiv:1303.6254
8. Czakon M, Mitov A (2014) Top++: a program for the calculation of the top-pair cross-section at hadron colliders. Comput Phys Commun 185:2930 arXiv:1112.5675 [hep-ph]
9. Botje M et al (2011) The PDF4LHC working group interim recommendations. arXiv:1101.0538 [hep-ph]
10. Martin A et al (2009) Parton distributions for the LHC. Eur Phys J C 63:189 arXiv:0901.0002 [hep-ph]
11. Martin A et al (2009) Uncertainties on α_s in global PDF analyses and implications for predicted hadronic cross sections. Eur Phys J C 64:653 arXiv:0905.3531 [hep-ph]
12. Lai H-L et al (2010) New parton distributions for collider physics. Phys Rev D 82:074024 arXiv:1007.2241 [hep-ph]
13. Ball RD et al (2013) Parton distributions with LHC data. Nucl Phys B 867:244 arXiv:1207.1303 [hep-ph]
14. Gao J et al (2014) CT10 next-to-next-to-leading order global analysis of QCD. Phys Rev D 89(3):03009 arXiv:1302.6246 [hep-ph]

15. Collaboration CMS (2018) Measurements of $t\bar{t}$ cross sections in association with b jets and inclusive jets and their ratio using dilepton final states in pp collisions at $\sqrt{s} = 13$ TeV. Phys Lett B 776:335 arXiv:1705.10141 [hep-ex]
16. Ježo T, Lindert JM, Moretti N, Pozzorini S (2018) New NLOPS predictions for $t\bar{t} + b$-jet production at the LHC. Eur Phys J C 78:502. arXiv:1802.00426 [hep-ph]
17. Garzelli MV, Kardos A, Trocsanyi Z (2015) Hadroproduction of $t\bar{t}b\bar{b}$ final states at LHC: predictions at NLO accuracy matched with Parton Shower. JHEP 03:083 arXiv:1408.0266 [hep-ph]
18. Bevilacqua G, Garzelli M, Kardos A (2017) $t\bar{t}b\bar{b}$ hadroproduction with massive bottom quarks with PowHel. arXiv:1709.06915 [hep-ph]
19. Debnath D et al (2017) Resolving combinatorial ambiguities in dilepton $t\bar{t}$ event topologies with constrained M_2 variables. Phys Rev D 96:076005 arXiv:1706.04995 [hep-ph]
20. Collaboration CMS (2012) Observation of a new boson at a mass of 125 GeV with the CMS experiment at the LHC. Phys Lett B 716:30 arXiv:1207.7235 [hep-ex]
21. Collaboration CMS (2019) Measurement of inclusive and differential Higgs boson production cross sections in the diphoton decay channel in proton-proton collisions at $\sqrt{s} = 13$ TeV. JHEP 01:183 arXiv:1807.03825 [hep-ex]
22. Collaboration CMS (2017) Measurements of properties of the Higgs boson decaying into the four-lepton final state in pp collisions at $\sqrt{s} = 13$ TeV. JHEP 11:047 arXiv:1706.09936 [hep-ex]
23. Collaboration CMS (2019) Search for $t\bar{t}H$ production in the $H \to b\bar{b}$ decay channel with leptonic $t\bar{t}$ decays in proton-proton collisions at $\sqrt{s} = 13$ TeV. JHEP 03:026 arXiv:1804.03682 [hep-ex]
24. Collaboration CMS (2015) Search for a standard model Higgs boson produced in association with a top-quark pair and decaying to bottom quarks using a matrix element method. Eur Phys J C 75:251 arXiv:1502.02485 [hep-ex]
25. Collaboration CMS (2014) Search for the associated production of the Higgs boson with a top-quark pair. JHEP 09:087 arXiv:1408.1682 [hep-ex]
26. Collaboration CMS (2013) Search for the standard model Higgs boson produced in association with a top-quark pair in pp collisions at the LHC. JHEP 05:145 arXiv:1303.0763 [hep-ex]
27. ATLAS Collaboration (2018) Search for the standard model Higgs boson produced in association with top quarks and decaying into a $b\bar{b}$ pair in pp collisions at $\sqrt{s} = 13$ TeV with the ATLAS detector. Phys Rev D 97:072016. arXiv:1712.08895 [hep-ex]
28. ATLAS Collaboration (2015) Search for the standard model higgs boson produced in association with top quarks and decaying into $b\bar{b}$ in pp collisions at $\sqrt{s} = 8$ TeV with the ATLAS detector. Eur Phys J C 75:349. arXiv:1503.05066 [hep-ex]
29. Cowan G, Cranmer K, Gross E, Vitells O (2001) Asymptotic formulae for likelihood-based tests of new physics. Eur Phys J C 71:1554 arXiv:1007.1727 [physics.data-an]
30. Read AL (2002) Presentation of search results: the CL$_s$ technique. J Phys G 28:2693
31. Bhat PC (2011) Multivariate analysis methods in particle physics. Ann Rev Nucl Part Sci 61:281
32. Drucker H, Cortes C (1995) Boosting decision trees. In: Proceedings of the 8th conference on advances in neural information processing systems (NIPS), Denver, USA, vol 8, p 479
33. Kondo K (1988) Dynamical likelihood method for reconstruction of events with missing momentum: I. method and toy models. J Phys Soc Jpn 57:4126
34. Kondo K (1991) Dynamical likelihood method for reconstruction of events with missing momentum: II. mass spectra for $2 \to 2$ processes. J Phys Soc Jpn 60:836
35. Dalitz RH, Goldstein GR (1992) Decay and polarization properties of the top quark. Phys Rev D 45:1531
36. CMS Collaboration (2005) CMS computing: technical design report. CERN-LHCC-2005-023. http://cds.cern.ch/record/838359
37. CMS Collaboration (2005) LHC computing grid: technical design report. CERN-LHCC-2005-024. http://cds.cern.ch/record/840543
38. Erdmann M et al (2018) The VISPA internet-platform in deep learning applications. J Phys: Conf Ser 1085:042044

Chapter 5
Analysis Technologies

The strategy and implementation of this analysis are based on a set of technological key components. They can be divided into three parts and are described in the following sections.

The first section discusses the analysis workflow management system (AWM[1]) that was developed alongside this analysis to cope with its scale and the complex relations among particular workloads. Subsequently, a summary of the main aspects of deep learning and neural networks is presented. It comprises all concepts that are applied during design and training of the neural networks as used for event categorization and process discrimination (cf. Sect. 4.2.2 and Chap. 7). The third section introduces the statistical framework that constitutes the basis of the extraction of results in terms of expected and observed limits, significances, and best fit values of the $t\bar{t}H$ signal strength modifier (cf. Sects. 4.2.1). In addition, it explains the goodness-of-fit quantity that is utilized to assess the agreement between recorded and simulated event variables to be used for neural network training.

5.1 Analysis Workflow Management

High-level physics analyses usually consist of a considerable amount of logically separated workloads. In general, the interface between these workloads does not rely on an event-by-event data flow such as utilized by, for example, CMSSW (cf. Sect. 3.4.1) or PXL (cf. Sect. 3.4.2). The workloads to perform a specific analysis often form a loose collection of inhomogeneous procedures, encoded in executable files such as Shell and Python scripts, and are executed manually. Hereby, their execution order

---

[1]The abbreviation AWM is favored over the common term WMS in the following, as the latter often refers to workload management systems in the context of remote job scheduling.

© The Author(s), under exclusive license to Springer Nature Switzerland AG 2021   85
M. Rieger, *Search for t̄tH Production in the H → bb̄ Decay Channel*, Springer Theses,
https://doi.org/10.1007/978-3-030-65380-4_5

is dictated by the dependencies between them in terms of the existence of results of one or more previous procedures. Beyond a certain scale and complexity, the manual steering of analysis workflows can be time-consuming, prone to errors, and potentially leads to undocumented relations between workloads.

A novel software for workflow management was developed alongside this analysis to cope with the challenges and risks of inhomogeneous workload definition and manual steering. It is based on the pipelining package LUIGI [1] (cf. Sect. 3.4.4) due to its simple yet scalable and extensible structure. To meet the sophisticated demands of performance-intensive analyses, the AWM implements the building blocks to seamlessly integrate remote resources as used in the high-energy physics community. The eventual goal of the software is to provide a versatile working environment for robust analyses on scalable and interchangeable resources, while immediately enabling analysis preservation.

The following section presents basic concepts of the developed software. Thereafter, the techniques of remote execution, remote storage, and environment sandboxing are explained, and the resulting prospects for analysis preservation are summarized. A previous version of the AWM is described in Ref. [2]. The software is publicly available and developed further in an open-source effort [3].

5.1.1   Basic Concepts

Conceptually, the developed AWM is independent of any experiment, specific use case, programming language, or data format. In typical high-energy physics nomenclature, it does not qualify as a "framework", but rather defines a collection of permissive guidelines and tools to follow a design pattern for distributed physics analyses. Specially tailored solutions already exist for a small number of particle physics applications such as the comprehensive event simulation campaigns performed centrally by the CMS and ATLAS experiments at the LHC [4, 5].

However, the requirement profile for performing end-to-end physics analyses can differ greatly. Particular workloads may require the results of multiple previously executed workloads and produce more than one output themselves. These dependencies form a directed acyclic graph (DAG), which can take arbitrary shapes and could even depend on dynamic run-time conditions. In addition, the shape of the DAG is not necessarily known a-priori and also might be subject to continuous development cycles. Furthermore, whereas central experiment workflows often rely on dedicated infrastructure, analyses must rather incorporate existing resources and maintain the ability to adapt to short-term changes.

The key operating principles of the AWM and the associated analysis design pattern are based on four paradigms which are discussed in the following paragraphs. A supporting illustration is presented in Fig. 5.1.

1. Encapsulation   Logically separated steps of an analysis are encapsulated in workloads. They can produce outputs, require the completion of certain other

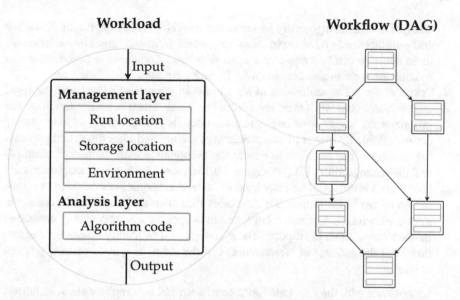

Fig. 5.1 Illustration of a workflow composed of multiple workloads. Workloads consist of a management layer that defines and handles run location, storage location, and environment, and an analysis layer comprising the actual algorithm code to run. The analysis layer is fully decoupled from the management layer, allowing for interchangeability of resources. Dependencies between workloads (arrows) result in a directed acyclic graph (DAG)

workloads, and therefore consider their outputs as potential inputs. Workloads and the dependencies between them compose a workflow that is visualizable as a DAG. The graph structure is scalable, versatile, and can be exploited by execution systems with automatic dependency resolution to process entire workflows using a single command invocation.

2. **Factorization**   In structural terms, a workload is factorized into four domains that are organized in two layers. The "management" layer handles the run location of a workload, defines the location where output data is stored, and steers the software environment during execution. The "analysis" layer contains the actual algorithm code to run. It is fully decoupled from the management layer, i.e., algorithms remain unaffected by any decisions about run and storage locations. Dynamic behavior of both layers can be achieved by including parameters to the workload. Their values might be configured at execution-time or depend on run-time conditions.

3. **Interchangeability**   Implementations of domains of the management layer are interchangeable. For instance, the change of the storage location for produced workload outputs neither affects the analysis algorithm, nor decisions about its run location or software environment. Evidently, exceptions occur in cases where special hardware requirements (e.g. GPUs for machine learning tasks), or strict dependencies between run and storage locations exist (e.g. data access on remote computing elements behind firewalls is often restricted to dedicated storage sys-

tems). The interchangeability of resources provides a high degree of flexibility and prevents hard-coded dependence on certain infrastructure. Decisions on all three domains can be made at the execution-time. Switching between local and remote resources is also beneficial for testing purposes.

4. **Universality**   The realization of algorithms in the analysis layer and the type, quantity, and content of their results are universal. There is no limitation on the programming language or implementation details of algorithms, as long as they are executable by means of sub-processing. Before and after the actual processing, measures can be taken to ensure the compliance with the factorization (2) and interchangeability (3) paradigms. Furthermore, the form of analysis results is entirely unrestricted, i.e., any kind of stateful information can serve as a definition of workload output. This includes files with arbitrary data formats, file system directories, database entries, and even specific file content, e.g. contained in machine-parsable job reports. The paradigm of universality causes the distinction from the concept of "frameworks" in the sense of typical particle physics applications.

Compliance with the four paradigms entails several accomplishments. Relations between workloads are encoded as dependencies and become integral parts of an analysis. Besides the evident advantages for automation, this also leads to notable benefits for the purpose of documentation. Beyond a certain scale and complexity, manual steering of physics analyses can be a challenging task. Especially in working groups where analyses are fully or partially operated by multiple individuals, a common interface between workloads has significant advantages. Final and intermediate results are both reproducible and reusable between similar analyses, effectively lowering the degree of unnecessary redundancy. Furthermore, the DAG defined by a workflow can be visualized with appropriate tools and, in environments such as provided by standard web browsers, can even serve as a basis for status monitoring during execution.

Technically, the concepts and paradigms introduced above are realizable using the scalable and extensible structure provided by the LUIGI package (cf. Sect. 3.4.4). The AWM is written in Python, which is a common choice in scientific applications owing to its high acceptance, support for diverse programming patterns like object orientation and scripting, and the vast amount of available libraries. Using LUIGI, workloads are described by parameterizable "tasks" that may require other tasks, define output "targets", and separately designate the algorithm code to run. Outputs of required tasks denote inputs, and a task is considered complete when all its output targets exist, independent of how the criterion of existence is evaluated. Details regarding implementation of the four paradigms as well as technical aspects are presented in the subsequent sections.

5.1.2 Remote Execution

Physics analyses investigating rare processes in high-energy particle collisions typically examine a considerable amount of data to infer statistically significant results. This implies that a large quantity of computing tasks must be executed, which most likely exceed the capacity of local host systems in terms of processing units, memory consumption, and storage space. The number of unique tasks to process in the scope of the presented analysis is, for example, in the order of $\sim 10^5$. Consequently, remote task execution on distributed batch systems is essential. Following the paradigm of factorization as introduced above, analysis algorithm code should remain unaffected by any decisions about the run location.

The developed AWM realizes remote execution capabilities through "mixin-based" inheritance. Tasks can inherit from one ore more classes that provide functionality for submitting computing jobs to certain remote batch systems as well as for retrieving job status information and results. In contrast to inheritance models that denote "is-a" relationships, mixin-based inheritance is considered as a technique to adapt common functionality while avoiding code redundancy. Thus, the configuration of run locations is rather declarative and, in particular, does not interfere with other factorized workload domains. In case of multiple available run locations provided by different mixin-classes, the actual implementation, i.e., the set of methods handling job submission and status retrieval for a particular batch system, can be selected at execution time using task parameters, which fulfills the demand for interchangeability. At the time of writing, the HTCondor [6, 7], LSF [8], ARC [9], and gLite [10] batch job interfaces are supported.

Another strict policy of the remote execution system is customizability. Deployed instances of the same batch system, such as HTCondor, can behave fairly different regarding submission requirements and overall supported features, depending on how they are configured by administrators. Therefore, the implemented execution system neither assumes a particular setup nor does it attempt to extrapolate specifications, but rather provides transparent access to all job configuration options. Furthermore, various features are included which are crucial for efficient operation on batch systems. Among others, they include job error detection, automatic resubmission, early stopping criteria, and the possibility to connect to common monitoring interfaces.

5.1.3 Remote Storage

The large amount of data to be processed inevitably causes the need for sophisticated data storage systems. Requirements on the storage capacity per analysis in the order of $\sim 10 - 100$ TB are not unusual. Especially when relying on remote execution systems that allow for a great number of parallel jobs such as the Worldwide LHC Computing Grid (WLCG) [11], the use of high-throughput storage is mandatory. Furthermore, storage locations should be accessible from within run locations in order to avoid manual copying of files and permanently occupying redundant disk space.

The concept of output targets, as introduced by the LUIGI package, represents a suitable abstraction for handling files on interchangeable remote resources. While multiple realizations of targets are conceivable in general (cf. Sect. 3.4.4), the following paragraphs focus solely on physical files. Independent of whether targets denote local or remote files, they strictly implement the same interface, which is defined by a common superclass and contains both low- and high-level file operations. Algorithms as part of the analysis layer can rely on the equivalence of operations, fully omitting distinction of cases in order to achieve interchangeability of storage locations. Additionally, the approach can be used to enable load distribution over several different locations. The respective behavior is configurable on target- and task-level.

The novel AWM provides a generic interface for remote file targets based on the open-source Grid File Access Library (GFAL) developed at CERN [12]. The library provides a Portable Operating System Interface (POSIX) for remote files accessible through most protocols employed in the landscape of high-energy physics such as provided by XRootD, dCache, SRM, and GridFTP, as well as further popular backends such as SFTP, WebDAV, Amazon S3, and Dropbox. Moreover, the developed target interface provides essential features including batch transfers, transfer validation, automatic retries for robustness against connection disruptions, and optional local caching. The latter is of particular importance for the optimization of network utilization in a variety of use cases. Examples are repeated local tests, or interdependent tasks running consecutively on the same machine and thus, being able to use cached versions of previously created output targets. Following the factorization paradigm, both cache synchronization and invalidation are fully integrated into all remote target operations and do not require special considerations in analysis algorithm code.

5.1.4   Environment Sandboxing

Another important aspect of the factorization paradigm is the management of software and environments. Different workloads within the same analysis workflow may depend on distinct software setups or even operating systems that are not necessarily compatible with each other. Examples are workloads that utilize key components of experiment-specific software that is often subject to extensive validation procedures and therefore requires long-term stability. This is opposed to machine learning algorithms built on top of third-party libraries, which may crucially improve in performance when updated on a regular basis. For this purpose, a mechanism referred to as environment "sandboxing" is integrated into the developed AWM.

A sandbox describes a collection of environment configurations, ranging from plain sets of variables, over to distinct software setups and different operating systems. Possible implementations encompass subshells with initialization files loaded on top of existing environments, virtual machines (VMs), or virtual environments

(VEs). VEs are particularly appealing as they are realized on the basis of container-ization, such as provided by Docker [13] or Singularity [14]. Tasks can declare the demand for a particular, favored sandbox and, optionally, provide further logic for negotiating with the current environment to determine if either compatibility is sufficient or sandboxing is necessary. In the latter case, the developed mechanism embeds the invocation of the algorithm code in the analysis layer within the spec-ified sandbox. This approach preserves the encapsulation paradigm by considering environment dependencies as part of the management layer of workloads and thus, enables the execution of inhomogeneous workflows with a single command.

Another benefit of sandboxing is related to long-term stability and portability. Most host systems and computing clusters are subject to constant changes owing to maintenance and security measures. For workloads with critical software depen-dencies, sandboxing provides a method for ensuring reproducibility of results over time. Therefore, it exhibits a key technique for the preservation of analyses, which is discussed in the following section.

5.1.5 Analysis Preservation

The full preservation of analyses is a key ingredient for sustainability of experimental particle physics research. Therefore, long-term storage of data, analysis algorithms, and related external information is required to ensure the repetition of measure-ments with reproducible results for the purposes of documentation, education, and reinterpretation. Especially reinterpretation campaigns represent an important use case as they could utilize preserved analyses to test hypotheses postulated by newly developed theories, exploiting the amount of recorded data and developed algorithms to this date. Inter-experimental collaborations have formed to advance this common effort [15]. In practice, however, the process proves to be challenging owing to the fact that analyses need to be adapted manually to fulfill certain preservation conditions. In contrast, analysis workflow management systems with integrated preservation capabilities could offer appealing benefits.

On the basis of the four paradigms introduced above (cf. Sect. 5.1.1), a decisive set of preservation conditions for workflow management systems can be derived.

1. The analysis layers of workloads containing algorithm implementations must not be changed. Moreover, operations relying on random quantities should be altered to use fixed seeds in random number generators.
2. Software and other relevant environment configurations must be retained such that repeated execution of the analysis code leads to identical results. External libraries might be updated when complying with prescriptions of semantic versioning.
3. All initial and external files must either be retained, or referenced to copies on dedicated data preservation systems.

4. For analyses that require a considerable amount of resources in terms of processing time, storage capacity, and specialized hardware, access to remote infrastructures providing parallel execution and storage systems must be configurable.
5. The execution of analysis workflows must be fully automated and / or sufficiently documented in order to guarantee operability by other researchers at any time.

The developed AWM intrinsically satisfies conditions 1, 4 and 5 by adhering to the paradigms of workload domain factorization and interchangeability of run and storage locations. The retention of software and environment configurations (condition 2) can be achieved by means of the sandboxing mechanism. Here, containerization methods such as provided by Docker [13] or Singularity [14] are particularly suitable. For high-energy physics experiments, the preservation of simulated and recorded collision events (condition 3) requires sophisticated measures, owing to the large amount of data in the order of $\sim\mathcal{O}(\text{PB})$. Access to extensive LHC open-data initiatives [16, 17] is mandatory and therefore supported by the AWM.

5.2   Deep Learning and Neural Networks

This section presents an overview of the key principles of multivariate analysis (MVA) with deep neural networks (DNN). As deep learning, and machine learning in general, constitutes a broad field of current research, which is subject to continuous development, the paragraphs below focus strictly on the concepts that are applied in this analysis. An extensive reference on deep learning and neural networks can be found in Ref. [18].

The sections below are structured as follows. First, the mathematical notation and construction of DNNs is introduced. After that, the process of neural network training is explained and the underlying minimization algorithm is discussed. The section closes with additional methods that target training optimization and model generalization.

5.2.1   Mathematical Construction

A neural network can be described as a function that, based on a well-structured mathematical model, maps an input x to an output y, both of which can have arbitrary dimensions and numeric domains. Depending on the particular use case, y denotes a "prediction" given the input "features" x in the context of a numerical regression or classification task. In general, the model consists of a composition of mathematical operations involving a vast amount (often $\gtrsim \mathcal{O}(10^6)$) of free parameters in order to exploit correlations and characteristic patterns in potentially highly complex input spaces. The determination of these parameters is subject to an optimization procedure, usually referred to as network "training", and consists of the minimization of

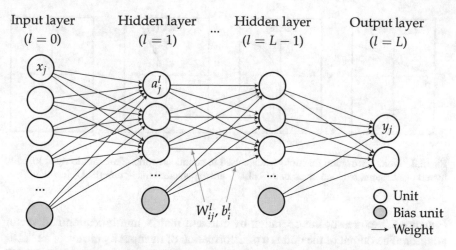

Fig. 5.2 Overview of a generic, fully-connected neural network with an input layer, $L - 1$ hidden layers, and an output layer. The input to each unit (white circles) is the weighted sum (arrows) of all units in the previous layer, while its output is subject to an activation function. Constant bias units ensure numeric independence of units in the next layer via additional weights

an objective function that compares predicted and true expected outputs to derive a "loss". Shape and behavior of this loss function are problem-specific and determined by the respective task to be accomplished. To ensure an effective and efficient training of the numerous free parameters, the neural network model is constructed in a layerwise fashion, which is explained in the following. The iterative training algorithm exploiting this structure is described subsequently.

Figure 5.2 shows a generic deep neural network with an input layer ($l = 0$), $L - 1$ "hidden" layers, and an output layer ($l = L$). Each layer consists of an adjustable number of "units", which receive values from all units in the previous layer, transform them into a single output value, and pass it to all units in the next layer. The number of input and output units x_j and y_j, respectively, are specified by the particular task, whereas the number and size of hidden layers are "hyperparameters" and subject to manual optimization apart from the training process. Arrows in Fig. 5.2 represent free parameters, or "weights" of the model and gray circles denote so-called "bias" units whose value is constant one. The latter enable numeric independence of units in the next layer as described below. Weights between unit j in layer l (a_j^l) and unit i in layer $l + 1$ (a_i^{l+1}) are denoted by W_{ij}^l, whereas weights from bias units to a_i^{l+1} are labeled as b_i^l.

The input z_i^l of a unit a_i^l is given by the weighted sum of the units in the previous layer,

$$z_i^l = \sum_j W_{ij}^{l-1} \cdot a_j^{l-1} + b_i^{l-1} \tag{5.1}$$

$$\Rightarrow z^l = W^{l-1} \cdot a^{l-1} + b^{l-1}. \tag{5.2}$$

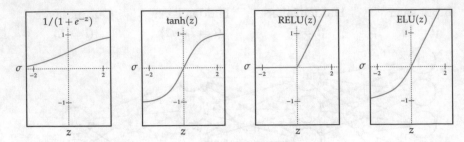

Fig. 5.3 Typical activation functions that are used in neural networks. From left to right: logistic, hyperbolic tangent, rectified linear unit (RELU), and exponential linear unit (ELU) function

Equation 5.2 expresses the operation by efficient matrix multiplication and vector addition. The output of the unit is a transformation of its input by means of an "activation" function σ^l. This transformation causes non-linear behavior of the model, which is essential for its ability to approximate arbitrary functional relations. A selection of common activation functions is illustrated in Fig. 5.3. All units of a particular layer are transformed with the same activation, whereas different functions can be chosen per hidden layer, representing additional hyperparameters. Thus, the output a of units in layer l can be written as

$$a^l = \sigma^l(z^l) = \sigma^l \left(W^{l-1} \cdot a^{l-1} + b^{l-1} \right) \tag{5.3}$$

$$\text{with } \sigma(z) := (\sigma(z_0), \ \sigma(z_1), \ \dots)^T . \tag{5.4}$$

It should be noted that Eq. 5.3 defines a recursive relation $a^l = a^l(a^{l-1})$. In general, the choice of the output layer activation depends strongly on the individual application. For multi-class classification tasks, a "softmax" transformation is used to scale the sum of the unit outputs to unity through

$$\sigma_{\text{softmax}}(z^L) = \frac{e^{z^L}}{\sum_i e^{z_i^L}}. \tag{5.5}$$

In conclusion, the model of a DNN can be expressed as a composition of its layers as

$$y = y(x \mid \underbrace{W, b}_{:= \omega}) = \underbrace{(a^L \circ a^{L-1} \circ \cdots \circ a^1)_\omega}_{\text{model}} (x), \tag{5.6}$$

given the set of free parameters ω. Their determination within the neural network training process is the subject of the subsequent section.

5.2.2 Neural Network Training

The procedure of neural network training commences with the definition of a task-specific loss function that compares the predicted output $y_n(x)$ of an example n to its true expected outcome $\hat{y}_n$. For the regression of a single quantity, the mean squared error (MSE) loss is a suitable choice, whereas (multi-class) classification tasks typically rely on the cross-entropy (CE) loss,

$$\text{Mean squared error:} \quad L(y(x), \hat{y})_{\text{MSE}} = \frac{1}{N} \sum_n^N (\hat{y}_n - y_n(x))^2 \quad (5.7)$$

$$\text{Cross-entropy:} \quad L(y(x), \hat{y})_{\text{CE}} = \frac{1}{N} \sum_n^N -\hat{y}_n \log y_n(x). \tag{5.8}$$

In case of the CE loss, $\hat{y}_n$ represents a vector whose length corresponds to the number of classes, with values of one denoting a specific class, while being zero otherwise. The logarithm is evaluated elementwise. The loss is often expressed as an average over a "batch" of N examples to obtain a statistically reliable quantity. For a fixed batch of examples, it can be reinterpreted as a function of the model weights, $L(y(x \mid \omega), \hat{y}) \rightarrow L(\omega \mid y(x), \hat{y})$. Then, the aim is to find a set of weights that minimizes the loss evaluated for a statistically representative amount of examples.

This minimization is achieved through an iterative, stochastic process referred to as batch gradient descent (BGD), which connects the change of the loss function to a rule for updating the model weights. For small variations, the change ΔL can be approximated as

$$\Delta L \approx \sum_i \frac{\partial L}{\partial \omega_i} \cdot \Delta \omega_i = \nabla_\omega L \cdot \Delta \omega. \tag{5.9}$$

By requiring that $\Delta \omega$ should be proportional to $\nabla_\omega L$, so that $\Delta \omega = -\eta \cdot \nabla_\omega L$, it follows

$$\Delta L \approx - \underbrace{\eta}_{>0} \cdot \underbrace{||\nabla_\omega L||^2}_{>0} < 0, \tag{5.10}$$

where the proportionality constant η is called the "learning rate". Thus, the loss change is negative and the total loss is minimized if the weights are adjusted according to the "update rule"

$$\omega_{t+1} = \omega_t - \eta \cdot \nabla_\omega L \tag{5.11}$$

after each training iteration t. However, the repeated calculation of the numerous and mostly complex derivatives $\nabla_\omega L$ is inefficient and computationally challenging.

A solution is provided by the so-called "backpropagation" algorithm [19] that exploits the layerwise structure of neural networks. It defines the "error" δ^l of a layer l as

$$\delta^l = \begin{cases} \nabla_y L \quad\quad\quad \odot\, \sigma'(z^l), & \text{for } l = L \\ \left(\left(W^{l+1}\right)^T \cdot \delta^{l+1}\right) \odot \sigma'(z^l), & \text{for } l < L \end{cases}, \tag{5.12}$$

where $\odot$ denotes the Hadamard product, i.e., elementwise multiplication, and $\sigma'(z^l)$ refers to the derivative of the activation function of layer l, evaluated at z^l. Therefore, the error of each unit in all network layers can be determined since $\nabla_y L$ and $\sigma'(z^l)$ are efficiently computable. In order to quantify $\nabla_\omega L$ for updating the model weights during training (Eq. 5.11), it can be shown [18, 19] that the derivatives are related to the errors δ (Eq. 5.12) through

$$\frac{\partial L}{\partial W_{ij}^l} = a_j^{l-1} \cdot \delta_i^l \qquad \text{and} \qquad \frac{\partial L}{\partial b_j^l} = \delta_j^l. \tag{5.13}$$

Thus, the weights can be updated efficiently and the full neural network training with backpropagation proceeds as follows.

1. Input the features x of a batch of N examples to the network.
2. Propagate unit outputs a^l in forward direction through all layers, compute the output y (Eq. 5.6), and determine the loss L (e.g. Eq. 5.7 or 5.8).
3. Determine the error δ^L in the last layer and propagate it back to obtain the errors of all previous layers δ^l (Eq. 5.12) and, subsequently, the derivatives $\nabla_\omega L$ (Eq. 5.13).
4. Update all weights according to the update rule (Eq. 5.11).

This procedure is repeated until the loss reaches a minimal value. As the number of available examples is usually limited, a special treatment is advisable for monitoring the network's ability of generalization. In scenarios where the set of training examples is an insufficient representation of the underlying, general probability distributions, or if the network's "capacity" is too extensive with respect to the number of available examples, the network might become sensitive to characteristic features of the training dataset. Therefore, an independent "validation" dataset should be retained to compute a validation loss in each iteration, which should be identical to the training loss used for backpropagation. In case of persistent deviations, the model fails to generalize, which is called "overtraining", the training should be stopped, and measures for overtraining suppression should be taken (cf. Sect. 5.2.3). As the validation dataset is therefore effectively utilized to optimize hyperparameters, an independent, third "testing" (or "evaluation") dataset is required for actual application in subsequent analyses.

5.2.3 Further Concepts

The implementation of the DNNs as used in this analysis employs further deep learning concepts that are discussed in the following paragraphs. They consist of methods for optimizing the training procedure and for suppressing overtraining.

Training Optimization

As explained in Sect. 5.2.2, weight updates during training crucially depend on the layerwise computed errors δ (Eqs. 5.11 and 5.13) and thus, on the gradients $\sigma'(z)$ of the activation functions, evaluated at the layer input z (Eq. 5.12). Gradients close to zero lead to negligible errors and, in turn, to vanishing weight updates, which impede the network's ability to learn over time. This "vanishing gradient" problem can be addressed with various approaches.

Since each input feature x_j may have a different numeric domain, the input to the first hidden layer, $z^1 = W^0 \cdot x + b^0$, is generally unrestricted. However, activation functions, such as shown in Fig. 5.3, tend to possess preferable regions, usually around zero, in which their values as well as derivatives are most meaningful in the context of backpropagation. As a consequence, input features and weights are subject to certain considerations. Firstly, each input x_j is transformed so that its distribution has a mean value of zero and a standard deviation of one,

$$x_j \rightarrow x'_j = \frac{x_j - \mu_j}{\sigma_j}. \tag{5.14}$$

Secondly, the initial weights should be drawn from (e.g.) a uniform or Gaussian distribution, whose parameters depend on the shape of W^0 and the chosen activation function in order to ensure an equally appropriate numeric domain of the layer output $a^1 = \sigma^1(z^1)$. Examples are the He [20] and the Glorot [21] initialization approaches. Bias weights are commonly initialized to zero, but updated during training to compensate for numeric offsets. In subsequent layers, weights can be initialized likewise, however, further methods exist to prevent vanishing gradients more efficiently, which especially occur in particularly deep networks. Batch normalization performs a scaling of hidden layer activations, similar to Eq. 5.14, with adaptive scaling coefficients per unit [22]. Moreover, the application of the scaled exponential linear unit (SELU) activation function with intrinsic self-normalizing behavior could be beneficial [23].

Besides adjusting input features and parameters of the network model, refinements of the loss minimization algorithm itself can be considered. One example is the adaptive moment estimation (ADAM) algorithm, which "computes individual adaptive learning rates for different parameters from estimates of first and second moments of the gradients" [24]. Per weight ω_i, it defines the linear (mean) and quadratic (variance) moments at a training iteration t as

$$m_{i,t} = \beta_1 \cdot m_{i,t-1} + (1 - \beta_1) \cdot \frac{\partial L}{\partial \omega_i} \qquad (5.15)$$

$$v_{i,t} = \beta_2 \cdot v_{i,t-1} + (1 - \beta_2) \cdot \frac{\partial L}{\partial \omega_i}^2, \qquad (5.16)$$

where $\beta_{1,2} \in [0, 1)$ represent exponential decay coefficients. While the gradients are obtained by the backpropagation algorithm as before (Eq. 5.13), the weight update rule is amended to

$$\omega_{i,t+1} = \omega_{i,t} - \eta \cdot \frac{1}{\sqrt{v_{i,t}/(1 - \beta_2^t)} + \epsilon} \cdot \frac{m_{i,t}}{1 - \beta_1^t}, \qquad (5.17)$$

with $\beta_{1,2}^t$ decreasing over time and a small number ϵ for numerical stability. If one imagines the loss function as a high-dimensional surface, the ADAM update rule can be perceived as an object descending along the surface with momentum and friction. The algorithm yields faster global convergence [24] and is less sensitive to perturbations due to local minima.

Overtraining Suppression

Overtraining occurs when the dataset used for training is an insufficient representation of the underlying, general probability distributions, either due to a bias in the selection of examples, or owing to a limited amount of statistics, while the capacity of the network is sufficiently large to learn characteristic, but non-representative features. Sophisticated applications require networks to find complex inner representations in data and hence, rely on an appropriate model that involves a considerable amount of free parameters. This trade-off situation is illustrated in Fig. 5.4. The neural networks

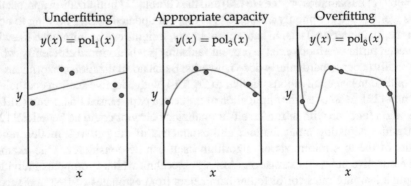

Fig. 5.4 Illustration of model capacity, under- and overfitting. Given a set of examples (x_i, y_i), a model with too few parameters is an insufficient representation of the data (left), while a too complex model is prone to fluctuations, might not generalize, and requires measures for suppressing overtraining (right). While a model with an appropriate capacity (center) appears reasonable, it might not be able to find complex inner representations of the input data

developed in this analysis employ two methods for overtraining suppression, which are described in the following two paragraphs.

In overfitting scenarios such as depicted in Fig. 5.4, weights of trained networks tend to take on large values, both positive and negative, to compensate for local fluctuations in a statistically non-representative input space. In order to retain the network's size and thus, its ability to model complex representations, the loss function can be extended to penalize large weights during training. This method, called L_2 regularization, adds a term

$$L_2 = \lambda \sum_i \omega_i^2 \tag{5.18}$$

to the loss function, which is mediated by a hyperparameter λ. Therefore, the training procedure minimizes the objective function defined by the task to be accomplished, while simultaneously avoiding excessively large weights. Bias weights are usually excluded in this approach.

Another approach for overtraining suppression is random unit dropout. It prevents networks from developing a too strong reliance on particular units during training but rather distribute the dependence among a statistically reliable amount of units. Given a customizable dropout rate $r \in [0, 1)$, unit outputs are randomly set to zero during training with a probability of r. As the average input z to the subsequent layer is reduced by a factor of $(1 - r)$, z is rescaled by a factor of $1/(1 - r)$ for compensation. For considerably large rates, the training with dropout can be understood as a training of an ensemble of networks [25].

5.3 Statistical Inference

In this section, the statistical framework for inferring the parameter of interest, namely the $t\bar{t}H$ signal strength modifier $\mu = \sigma/\sigma_{SM}$, is presented. As explained in Sect. 4.2.1, a signal extraction procedure is performed simultaneously in all analysis categories through a binned maximum likelihood fit of a simulated, parametrized variable distribution to recorded data. While μ is treated as an unconstrained model parameter, systematic uncertainties are incorporated as nuisance parameters involving a-priori assumptions derived from knowledge on theoretical cross sections, detector effects, and analytical methods. The construction of the underlying statistical model as a likelihood is described in the beginning.

Thereafter, the methods required for the inference of results in the context of a search for a new physics process are introduced. They comprise the determination of an upper limit on μ at 95% confidence following a modified frequentist CL_S approach, the exclusion significance of the background-only hypothesis, and the profile likelihood fit for the extraction of μ.

The section closes with a description of the χ^2-based goodness-of-fit test using a saturated model. The test is employed to quantify the agreement of particular variable correlations between simulated and measured events. A thorough validation is especially necessary for variables that are potentially considered as inputs to the deep neural networks developed in this analysis (cf. Sect. 7.1).

The technical realization of the statistical model and the signal extraction procedure strictly follows LHC recommendations for analyses in the Higgs sector [26]. The inference methods are performed with the so-called `combine` tool in version 7.0.1, developed and used by the CMS experiment [27].

5.3.1 Likelihood Construction

The construction of the likelihood can be motivated by considering a counting experiment. The probability to count a number of x events follows a Poisson distribution

$$\text{Poisson}(x \mid \lambda) = \frac{\lambda^x}{x!} e^{-\lambda} \tag{5.19}$$

with the expected yield of events λ. This approach can be extended to multiple bins N by multiplication of their Poisson probabilities,

$$p(x \mid \lambda) = \prod_{i}^{N} \text{Poisson}(x_i \mid \lambda_i), \tag{5.20}$$

where x_i and λ_i are the measured and expected event counts, respectively, in bin i. In practice, the set of N bins comprises all histogram bins of a selected variable distribution in several measurement categories (cf. Sect. 4.2.1).

$p(x \mid \lambda)$ denotes the conditional probability to measure x events given a particular statistical model. In the context of high-energy physics analyses, this model contains our knowledge about production mechanisms and rates of certain processes, and the amount of recorded and selected collision events in terms of luminosity, detector acceptance, and reconstruction efficiencies. Uncertainties of the model, both systematical and statistical, are incorporated as "nuisance" parameters θ [28] that are, in Bayesian terms, subject to prior probability distributions, often estimated through theoretical reasoning or external measurements. Thus, the number of expected events in bin i can be written as

$$\lambda_i = \mu \cdot s_i(\theta) + b_i(\theta), \tag{5.21}$$

with the numbers of signal and background events s_i and b_i, respectively. Except for the influence of certain nuisances, the signal yield enters Eq. 5.21 unconstrained and is factorized by the parameter of interest μ, which is not attributed any a-priori assumptions. Its prior probability distribution is said to be "flat".

For a specific, fixed measurement x, the likelihood as a function of the model parameters (μ, θ) can be formulated as [29, 30]

$$L(\mu, \boldsymbol{\theta} \mid \boldsymbol{x}) := p(\boldsymbol{x} \mid \mu, \boldsymbol{\theta}) = \prod_i^{\text{bins}} \text{Poisson} (x_i \mid \mu \cdot s_i(\boldsymbol{\theta}) + b_i(\boldsymbol{\theta})) \,. \qquad (5.22)$$

In general, the nuisances $\boldsymbol{\theta}$ may also depend on the bin i as well as on particular processes that contribute to $b_i(\boldsymbol{\theta})$. This dependence is implemented as a composition

$$b_i(\boldsymbol{\theta}) = \sum_p^{\text{processes}} \underbrace{b_{p,i}}_{\substack{\text{nominal} \\ \text{yield}}} \cdot \underbrace{\prod_n^{\dim(\boldsymbol{\theta})} \pi_{n,p,i}(\theta_n)}_{\substack{\text{uncertainty} \\ \text{model}}}, \qquad (5.23)$$

which reads analogously for $s_i(\boldsymbol{\theta})$. $b_{p,i}$ denotes the nominal, expected number of events in bin i contributed by process p, and $\pi_{n,p,i}(\theta_n)$ is the prior probability distribution of nuisance n evaluated at θ_n. The multiplicative term in the right part of Eq. 5.23 is restricted to positive values and can be understood as an uncertainty model.

Since not all nuisances act on all processes and bins, some $\pi_{n,p,i}$ are constant one. Others apply to all bins of a process ($\pi_{n,p,i} \rightarrow \pi_{n,p}$), all processes in a bin ($\pi_{n,p,i} \rightarrow \pi_{n,i}$), or even both ($\pi_{n,p,i} \rightarrow \pi_n$), such as the nuisance that describes the uncertainty on the measurement of the integrated luminosity. All prior distribution functions are continuous and defined such that $\pi_{n,p,i}(\theta_n) \geq 0 \; \forall \; \theta_n$. In this analysis, three types of nuisances are distinguished regarding the shape of their prior distributions and correlated impact on processes and bins.

1. **Rate-changing nuisances**   Some uncertainties, such as the uncertainties on theoretical cross sections, affect the overall normalization of one or more processes. This relation can apply for all bins considered in the analysis, or it might be limited to a subset of bins, referred to as a "template" and usually defined by a measurement category. Up and down variations of these uncertainties, describing central 68% confidence intervals, do not lead to shape changes of a template, but rather vary the total number of expected events, i.e., the integral of the template, equally in all its bins. Corresponding nuisances involve normal Gaussian prior probability distributions with mean one to predict the impact on the event rate of the uncertainty depending on the likelihood of the nuisance parameter θ_n. For large uncertainties, however, the normal distribution is rather broad and thus, must be truncated at zero in order to avoid negative values when computing the logarithm of the likelihood function [28]. Such a truncation could lead to systematically overestimated event yields. Therefore, a log-normal distribution is employed instead and parametrized accordingly as

$$\pi(\theta) = \frac{1}{\theta} \cdot \frac{1}{\sqrt{2\pi}\sigma} \exp\left[-\frac{(\ln\theta)^2}{2\sigma^2} \right]. \qquad (5.24)$$

Here, σ is the width of the distribution and corresponds to the relative uncertainty, which is estimated a-priori on the basis of theoretical calculations or external measurements.

2. **Shape-changing nuisances**   As opposed to uncertainties that only affect template normalizations, some uncertainties introduce variations of expected event yields that are correlated among bins of a template. Methodically, these shape-changing variations are produced by recreating a template within the analysis, where the model parameter in question is varied according to the boundaries of its central 68% confidence interval. Besides the nominal prediction, this results in two variants of the template that characterize the change of its shape given a discrete variation of the underlying parameter by ± 1 standard deviation. The applicability within the likelihood function, however, requires a continuous relation between the value of a nuisance θ and the impact on the template.

This relation is provided by the template "morphing" procedure, which is conducted independently in each bin following a cubic-linear approach [31]. Within the $\pm 1\,\sigma$ interval, the morphing algorithm interpolates between the central and the two varied templates using a polynomial of third degree, whereas a linear extrapolation is applied outside that region. More details on the cubic-linear approach are presented in Ref. [32]. Fig. 5.5 depicts different morphing scenarios. Owing to a potentially insufficient number of events in a nominal or varied template bin (scenario 1), the morphing function might predict negative yields for certain nuisance values. These occurrences are expected to vanish for statistically significant bin contents, which is addressed by a dedicated bin-merging algorithm (cf. Sect. 7.3). In general, the normalization of a template is not conserved during morphing and thus, it might have a rate- and shape-changing effect. However, by rescaling each

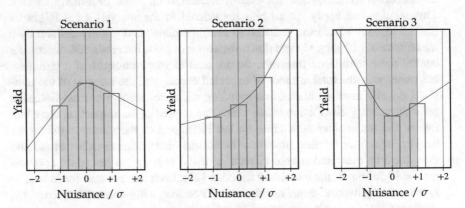

Fig. 5.5 Illustration of different template morphing scenarios. Values at zero represent the nominal event yield whereas $\pm 1\,\sigma$ variations denote shifts due to a nuisance, corresponding to one standard deviation. In the central region (gray), the morphing function is approximated by a polynomial of third degree, while a linear extrapolation is used outside (white). Owing to potentially insufficient bin statistics, the morphing function might describe negative yields in outer regions (scenario 1), which is to be avoided by choosing appropriate bin widths

bin by the ratio of integrals before and after applying the θ-dependent morphing procedure, the normalization is optionally retained, effectively leading to a solely shape-changing behavior.

3. **Statistical uncertainties** The amount of available statistics of simulated events that estimate the expected yield for each process and bin is limited. One possibility to model statistical uncertainties is to introduce one separate nuisance per process and bin with a prior probability following a Poisson distribution [28]. However, for a sufficiently large number of simulated events per bin, the sum of Poisson distributions can be approximated by a single Gaussian distribution with reasonable accuracy [33]. As a result, one nuisance per bin is incorporated in the statistical model. The precondition on the minimal bin content is met by means of a binning optimization (cf. Sect. 7.3).

Instead of performing a maximization of the likelihood function as constructed in this section to infer estimates for model parameters and the parameter of interest, it is advisable and advantageous to define a robust test statistic. Therefore, the following section motivates the concept of likelihood ratios and their utilization in statistical inference methods.

5.3.2 Inference Methods

Statistical inference methods in the context of searches for new physics processes are often predicated upon a hypothetical signal that might exist in a well-defined phase space in addition to a known background. Assuming the absence of potential interference effects, one defines a null hypothesis describing the non-existence of the signal process ($\mu = 0$), referred to as "background-only hypothesis" in the following, and an alternative hypothesis postulating the presence of signal ($\mu > 0$), called "signal hypothesis". Independent of the specific signal model, a discovery is usually claimed when the measured data is incompatible with the background-only hypothesis.

Following the Neyman–Pearson lemma [34], the most powerful test statistic Q to evaluate two contrary hypotheses is given by their likelihood ratio. Let κ_α be the threshold of a test statistic below which the null hypothesis $\theta = \theta_0$ is accepted. Then, for a fixed measurement x, the probability to falsely reject the null hypothesis is

$$\alpha = P(Q(x) \geq \kappa_\alpha \,|\, \theta = \theta_0), \tag{5.25}$$

whereas the probability to falsely accept it over the alternative hypothesis $\theta = \theta_1$ is given by

$$\beta = P(Q(x) < \kappa_\alpha \,|\, \theta = \theta_1). \tag{5.26}$$

The lemma states that the power of the test, $1 - \beta$, is maximized when Q is the likelihood ratio. As it originally does not apply to probability models with multiple free parameters, a generalized approach can be deduced on the basis of likelihood "profiling", which is explained in the subsequent paragraphs.

The profile likelihood ratio for the likelihood introduced in Eq. 5.22 is defined as [35]

$$\lambda(\mu) = \frac{L(\mu, \hat{\hat{\boldsymbol{\theta}}}(\mu))}{L(\hat{\mu}, \hat{\boldsymbol{\theta}})}.$$ (5.27)

It should be noted that λ implicitly depends on the measurement x, which is omitted in the nomenclature below for simplicity. $\hat{\mu}$ and $\hat{\boldsymbol{\theta}}$ in the denominator are the maximum likelihood estimators (MLE), i.e., the parameters that maximize L and minimize $-\ln(L)$, respectively. $\hat{\hat{\boldsymbol{\theta}}}(\mu)$ in the numerator denotes the conditional MLE that would maximize L for a prespecified value of the parameter of interest μ. Hence, the likelihood ratio λ solely depends on μ and is, in particular, independent of the nuisance parameters $\boldsymbol{\theta}$. This construction is called "profiling".

In case of a statistical deficit of signal-like events in a measurement x, $\hat{\mu}$, and likewise μ, can become negative. In order to avoid an artificial constraint on μ being positive, which would lead to calculatory complications [35], the test statistic is adjusted to

$$\widetilde{\lambda}(\mu) = \begin{cases} \frac{L(\mu, \hat{\hat{\boldsymbol{\theta}}}(\mu))}{L(\hat{\mu}, \hat{\boldsymbol{\theta}})}, & \hat{\mu} \geq 0 \\ \frac{L(\mu, \hat{\hat{\boldsymbol{\theta}}}(\mu))}{L(0, \hat{\hat{\boldsymbol{\theta}}}(0))}, & \hat{\mu} < 0 \end{cases},$$ (5.28)

which is equivalent to requiring $\mu \geq 0$. According to Wilks' theorem [36] and the generalization through Wald's theorem [37], if certain regularity conditions are met and the sample size is sufficiently large, the transformed test statistic

$$\widetilde{q}(\mu) = -2\ln\widetilde{\lambda}(\mu)$$ (5.29)

asymptotically converges towards a non-central χ^2 distribution with one degree of freedom [35]. It constitutes the basis for the following inference methods.

Upper Limit
The test statistic in Eq. 5.29 can be used to differentiate the hypothesis that a signal process is produced at a certain rate μ from the alternative stating that the production rate is smaller. This can be reflected by a case distinction

$$\widetilde{q}_\mu = \begin{cases} -2\ln\widetilde{\lambda}(\mu), & \hat{\mu} \leq \mu \\ 0, & \hat{\mu} > \mu \end{cases}.$$ (5.30)

For a given measurement x, a single value of the test statistic $\widetilde{q}_{\text{obs}}$ is obtained. To draw conclusions about the probability of measuring a value of $\widetilde{q}_{\text{obs}}$ or greater, its underlying probability distribution $f(\widetilde{q}_\mu \,|\, \mu)$ is required. As $\widetilde{q}$ is constructed from the profile likelihood ratio, it is independent of the nuisance parameters $\boldsymbol{\theta}$ in the asymptotic limit following Wilks' and Wald's theorems [36, 37], so that the conditional MLE $\hat{\hat{\boldsymbol{\theta}}}(\mu)$ provides a good estimate for the determination of $f(\widetilde{q}_\mu \,|\, \mu)$ [35]. This can be achieved by sampling from the full statistical model to create a "toy" dataset, or by using a single "Asimov" dataset. The latter represents an approximate dataset in which all free parameters are set to their expected values from simulation, giving rise to the non-centrality parameter of the χ^2 distribution that $\widetilde{q}$ converges to [35].

After determining $f(\widetilde{q}_\mu \mid \mu)$, the p-value describing the compatibility with the hypothetical signal strength μ is

$$CL_{\text{S+B}} = \int_{\widetilde{q}_{\text{obs}}}^{\infty} f(\widetilde{q}_\mu \mid \mu) \, d\widetilde{q}_\mu. \tag{5.31}$$

The frequentist upper limit on μ at 95% confidence could be defined by solving $CL_{\text{S+B}} = 0.05$ for μ. However, in case of statistical downward fluctuations, this upper limit would become arbitrarily small without being able to exclude $\mu = 0$ [38, 39]. Therefore, one defines the ratio

$$CL_{\text{S}} = \frac{CL_{\text{S+B}}}{CL_{\text{B}}}, \tag{5.32}$$

where the denominator is the p-value of the consistency with the background-only hypothesis,

$$CL_{\text{B}} = \int_{\widetilde{q}_{\text{obs}}}^{\infty} f(\widetilde{q}_\mu \mid 0) \, d\widetilde{q}_\mu. \tag{5.33}$$

As a result, the value of μ, obtained by setting $CL_{\text{S}} = 0.05$, denotes the observed, frequentist upper limit above which the signal strength is excluded by the measurement at 95% confidence. For blinded analyses, two variants of upper limits are commonly used. The "injected" limit exactly follows the above prescription, except that measured data is replaced with simulation. In contrast, for the calculation of the "expected" limit, $f(\widetilde{q} \mid \mu)$ is sampled from a statistical model that assumes the absence of signal. A value larger than one states that the background alone could explain the measurement, regardless of whether or not a signal exists. Therefore, the expected limit constitutes a sensitivity measure for blinded analyses.

Exclusion Significance

The claim of an observation is predicated on the exclusion of the background-only hypothesis $\mu = 0$. Hence, the test statistic is redefined to [35]

$$\widetilde{q}_0 = \begin{cases} -2 \ln \widetilde{\lambda}(\mu), & \hat{\mu} > 0 \\ 0, & \hat{\mu} \leq 0 \end{cases} \tag{5.34}$$

and the p-value describing the probability of the measurement assuming $\mu = 0$ reads similar to Eq. 5.33 as [35]

$$p_0 = \int_{\widetilde{q}_{\text{obs}}}^{\infty} f(\widetilde{q}_0 \mid 0) \, d\widetilde{q}_0. \tag{5.35}$$

Irrespective of whether the test statistic is normally distributed, the exclusion significance is often expressed in units of Gaussian standard deviations, i.e.,

$$p_0 = \int_s^\infty \frac{\exp(-x^2/2)}{\sqrt{2\pi}} \, dx \tag{5.36}$$

solved for s. A significance greater $s = 3$ ($p_0 = 1.3 \times 10^{-3}$) is called an "evidence" for the existence of a new physics process, while $s = 5$ ($p_0 = 2.9 \times 10^{-7}$) marks the threshold for claiming an "observation".

Profile Likelihood Fit

For the determination of the parameter of interest μ under the signal hypothesis, one minimizes the negative log-likelihood (NLL) $-2 \ln L(\mu, \boldsymbol{\theta} \mid \boldsymbol{x})$ [30], which is typically normalized by $L(\hat{\mu}, \hat{\boldsymbol{\theta}})$ (Eq. 5.27). In physics jargon, this minimization is often described as being a fitting procedure, with the MLE of μ constituting the "best fit value".

To simplify the process in case of multiple substantial nuisances, parameter profiling can be applied. For a fixed value $\mu = \mu'$, the nuisance parameter values that would minimize the NLL, i.e., the conditional MLEs $\hat{\hat{\boldsymbol{\theta}}}(\mu')$, are determined and the corresponding minimum of the NLL is retained. This procedure is repeated for different values of μ' and represents a scan over possible values of μ, where the overall minimal NLL value yields the best fit value $\hat{\mu}$,

$$\hat{\mu} = \arg \min_\mu -2 \ln L(\mu, \hat{\hat{\boldsymbol{\theta}}}(\mu) \mid \boldsymbol{x}). \tag{5.37}$$

An exemplary profile likelihood scan is shown in Fig. 5.6. The minimum of the resulting curve is shifted to zero as one is only interested in its position in terms of the parameter of interest. Therefore, and owing to the asymptotic χ^2-like behavior as discussed above, the generally asymmetric uncertainty on the best fit value is extracted through the intersection of the scanned curve with $-2\Delta \ln L = 1$.

Likewise, the best fit values and uncertainties of arbitrary nuisance parameters can be calculated by hypothetically considering them as the parameter of interest in the scan and profiling out all other parameters, while optionally fixing the actual parameter of interest μ to its best fit value. An alternative, but computationally more demanding approach is the scan in multiple dimensions. The resulting a-posteriori, or "post-fit" nuisances θ_{post}, and especially nuisance pulls, i.e., the change of central nuisance values and their probability distribution widths $\Delta\theta$ with respect to the a-priori, or "pre-fit" expectation θ_{pre},

$$\text{pull} := \frac{\theta_{\text{post}} - \theta_{\text{pre}}}{\Delta\theta_{\text{pre}}} \tag{5.38}$$

provide valuable insights to assess the validity of the underlying statistical model.

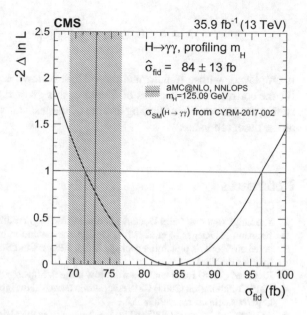

Fig. 5.6 Example of a shifted negative log-likelihood (NLL) distribution of the measurement of inclusive and differential Higgs boson production cross sections in the $\gamma\gamma$ final state performed by the CMS experiment [40]. The minimum is shifted to zero so that the uncertainty on the best fit value $\hat{\sigma}_{\text{fid}}$ is defined through the intersections with NLL = 1

5.3.3 Saturated Goodness-of-Fit Test

A goodness-of-fit (GOF) test describes the test of a hypothesis in the absence of a specific alternative. As a consequence, there is no single, optimal prescription for performing GOF tests, such as the Neyman–Pearson lemma for the evaluation of two contrary hypotheses [34]. The presented analysis employs the concept of binwise GOF tests to verify the modeling of correlations between variables that are considered as inputs to deep neural networks (cf. Sect. 7.1).

For sufficiently large statistics, the event yields per bin are normally distributed and the likelihood for a mean μ_i and variance σ_i^2 in bin i, given a measurement x_i is [41]

$$L(\boldsymbol{\mu}, \boldsymbol{\sigma} \mid \boldsymbol{x}) = \prod_i^{\text{bins}} \frac{1}{\sqrt{2\pi\sigma_i^2}} \cdot \exp\left(\frac{-(x_i - \mu_i)^2}{2\sigma_i^2}\right). \tag{5.39}$$

Now, the likelihood of the so-called "saturated" model is composed by setting all parameters μ_i to the measured values x_i [42] such that an artificial alternative hypothesis is formulated as

$$L_{\text{saturated}}(\boldsymbol{\mu} = \boldsymbol{x}, \boldsymbol{\sigma} \mid \boldsymbol{x}) = \prod_i^{\text{bins}} \frac{1}{\sqrt{2\pi\sigma_i^2}}. \tag{5.40}$$

Analogous to the construction of hypothesis tests in Sect. 5.3.2, the ratio

$$\lambda = L \,/\, L_{\text{saturated}} = \prod_i^{\text{bins}} \exp\left(\frac{-(x_i - \mu_i)^2}{2\sigma_i^2}\right) \qquad (5.41)$$

is employed, whose transformation $-2\ln\lambda$ follows a χ^2 distribution. Eventually, for the corresponding degrees of freedom, this gives rise to the p-value of the test, stating the probability of finding a value of the test statistic that is at least as extreme as the observed value.

References

1. The Luigi Authors. Luigi Documentation. https://luigi.readthedocs.io
2. Erdmann M, Rieger M et al (2017) Design and execution of make-like, distributed analyses based on Spotify's pipelining package Luigi. J Phys: Conf Ser 898:072047. arXiv:1706.00955 [physics.data-an]
3. Rieger M (2018) Luigi analysis workflow project. https://github.com/riga/law
4. CMS Collaboration (2005) CMS computing: technical design report. CERN-LHCC-2005-023. http://cds.cern.ch/record/838359
5. ATLAS Collaboration (2005) ATLAS computing: technical design report. CERN-LHCC-2005-022. http://cds.cern.ch/record/837738
6. Tannenbaum T, Wright D, Miller K, Livny M (2002) Beowulf cluster computing with linux: condor - a distributed job scheduler. MIT Press, Cambridge. ISBN 0262692740
7. Thain D, Tannenbaum T, Livny M (2005) Distributed computing in practice: the condor experience. Concurr Comput: Pract Exp 17:323
8. Lumb I, Smith C (2004) Grid resource management: scheduling attributes and platform LSF. Kluwer Academic Publishers, Norwell. ISBN 1402075758
9. Ellert M et al (2007) Advanced resource connector middleware for lightweight computational grids. Future Gener Comput Syst 23:219
10. Andreetto P et al (2008) The gLite workload management system. J Phys: Conf Ser119:062007
11. CMS Collaboration (2005) LHC computing grid: technical design report. CERN-LHCC-2005-024. http://cds.cern.ch/record/840543
12. Ayllon AA et al (2017) Making the most of cloud storage - a toolkit for exploitation by WLCG experiments. J Phys: Conf Ser 898:062027. http://cds.cern.ch/record/2297053
13. Boettiger C (2015) An introduction to Docker for reproducible research, with examples from the R environment. ACM SIGOPS Oper Syst Rev, Spec Issue Repeatability Shar Exp Artifacts 49:71 arXiv:1410.0846 [cs.SE]
14. Kurtzer GM, Sochat V, Bauer MW (2017) Singularity: scientific containers for mobility of compute. PLoS ONE 12:e0177459
15. DPHEP Collaboration (2012) Status report of the dphep study group: towards a global effort for sustainable data preservation in high energy physics. DPHEP-2012-001. arXiv:1205.4667 [hep-ex], http://cds.cern.ch/record/1450816
16. Cowton J et al (2015) Open data and data analysis preservation services for LHC experiments. J Phys: Conf Ser 664:032030. http://cds.cern.ch/record/2134548
17. Calderon A et al (2015) Open access to high-level data and analysis tools in the CMS experiment at the LHC. J Phys: Conf Ser 664:032027. http://cds.cern.ch/record/2015410
18. Goodfellow I, Bengio Y, Courville A (2016) Deep learning. MIT Press, Cambridge. ISBN 9780262035613, http://www.deeplearningbook.org
19. Rumelhart DE, Hinton GE, Williams RJ (1986) Learning representations by back-propagating errors. Nature 323:533

20. He K, Zhang X, Ren S, Sun J (2015) Delving deep into rectifiers: surpassing human-level performance on imagenet classification. In: Proceedings of the IEEE international conference on computer vision, Santiago, Chile
21. Glorot X, Bengio Y (2010) Understanding the difficulty of training deep feedforward neural networks. In: Proceedings of the 13th international conference on artificial intelligence and statistics, Sardinia, Italy, vol 9, p 249
22. Ioffe S, Szegedy C (2015) Batch normalization: accelerating deep network training by reducing internal covariate shift. In: Proceedings of the 32nd international conference on machine learning, Lille, France, vol 37, p 448. arXiv:1502.03167 [cs.LG]
23. Klambauer G, Unterthiner T, Mayr A, Hochreiter S (2017) Self-normalizing neural networks. Adv Neural Inf Process Syst 30. arXiv:1706.02515 [cs.LG]
24. Kingma DP, Ba JL (2015) Adam: a method for stochastic optimization. In: Proceedings of the 3rd international conference for learning representations, San Diego, CA, USA. arXiv:1412.6980 [cs.LG]
25. Hara K, Saitoh D, Shouno H (2016) Analysis of dropout learning regarded as ensemble learning. In: Proceedings of the 25th international conference on artificial neural networks, Barcelona, Spain, published in Springer LNCS, vol 2, p 1. arXiv:1706.06859 [cs.LG]
26. CMS and ATLAS Collaborations and LHC Higgs Combination Group (2011) Procedure for the LHC Higgs boson search combination in Summer 2011. CMS-NOTE-2011-005, ATL-PHYS-PUB-2011-011. http://cds.cern.ch/record/1379837
27. The CMS Higgs combine tool developers. https://cms-hcomb.gitbooks.io/combine
28. Conway JS (2011) Incorporating nuisance parameters in likelihoods for multisource spectra. In: Proceedings of the workshop on statistical issues related to discovery claims in search experiments and unfolding, CERN, Geneva, Switzerland. CERN–2011–006.115. arXiv:1103.0354 [physics.data-an]
29. James F (2006) Statistical methods in experimental physics. World Scientific Publishing, Singapore. ISBN 9789812567956
30. Baker S, Cousins RD (1984) Clarification of the use of chi-square and likelihood functions in fits to histograms. Nucl Instrum Meth 221:437
31. Baak M, Gadatsch S, Harrington R, Verkerke W (2014) Interpolation between multi-dimensional histograms using a new non-linear moment morphing method. CERN-OPEN-2014-050. arXiv:1410.7388 [physics.data-an]
32. Steggemann J (2012) Search for new particles decaying to a top quark pair with the CMS experiment. PhD Thesis, RWTH Aachen University. http://publications.rwth-aachen.de/record/197567
33. Barlow RJ, Beeston C (1993) Fitting using finite Monte Carlo samples. Comput Phys Commun 77:219
34. Neyman J, Pearson ES (1933) On the problem of the most efficient tests of statistical hypotheses. Phil Trans R Soc A 231:289
35. Cowan G, Cranmer K, Gross E, Vitells O (2001) Asymptotic formulae for likelihood-based tests of new physics. Eur Phys J C 71:1554 arXiv:1007.1727 [physics.data-an]
36. Wilks SS (1938) The large-sample distribution of the likelihood ratio for testing composite hypotheses. Ann Math Statist 9:60
37. Wald A (1943) Tests of statistical hypotheses concerning several parameters when the number of observations is large. Trans Amer Math Soc 54:426
38. Junk T (1999) Confidence level computation for combining searches with small statistics. Nucl Instrum Meth A 434:435 arXiv:hep-ex/9902006
39. Read AL (2002) Presentation of search results: the CL_s technique. J Phys G 28:2693
40. Collaboration CMS (2019) Measurement of inclusive and differential Higgs boson production cross sections in the diphoton decay channel in proton-proton collisions at $\sqrt{s} = 13$ TeV. JHEP 01:183 arXiv:1807.03825 [hep-ex]
41. D'Agostino RB, Stephens MA (1986) Goodness-of-fit-techniques. CRC Press, Boca Raton. ISBN 9780824774875
42. Lindsey JK (1996) Parametric statistical inference. Oxford Science Publications, Clarendon Press, Oxford. ISBN 0198523599

Chapter 6
Event Samples and Selection

This section introduces the samples of measured and simulated events that constitute the basis for the analysis at hand. Firstly, the sample of events and corresponding integrated luminosities measured with the CMS detector in the data-taking period of 2016 are described. Subsequently, the procedure of event simulation at the CMS experiment is summarized, the sample of simulated "Monte Carlo" (MC) events is introduced, and a detailed overview of the employed generator setup is presented. As contributions from $t\bar{t}$ processes with additional heavy-flavor jets constitute the most important background to this analysis (cf. Sects. 2.2.4 and 4.1), dedicated, heavy-flavor enriched $t\bar{t}$ event samples are included to increase the number of background events, which is especially beneficial for the training of machine learning algorithms (cf. Chap. 7). Thus, the second section closes with a description of the method for partially extending the statistics of the inclusive $t\bar{t}$ sample. The third section presents the selection criteria used to define the analysis phase space, divided into the single-lepton channel comprising events with either one electron or muon, and the dilepton channel with two oppositely charged electrons or muons. The subsequent section explains corrections that are applied to simulated events to account for effects that cause deviations between data and simulation, for example, regarding lepton identification or b-tagging efficiencies. Lastly, distributions of event observables are compared between data and simulation in all channels, and conclusions about the quality of the background modeling are drawn.

6.1 Recorded Data Event Samples

During the data-taking period of 2016, the LHC provided proton-proton collisions with a bunch spacing of 25 ns at a center-of-mass energy of $\sqrt{s} = 13$ TeV, reaching an instantaneous peak luminosity of $L = 0.77 \times 10^{34}$ cm^{-2}s^{-1} [1] (cf. Sect. 3.1). From a total integrated luminosity of 40.82 fb^{-1}, the CMS detector was able to record an amount of 37.76 fb^{-1} [2] with an overall uncertainty on the luminosity measurement of 2.5% [3].

© The Author(s), under exclusive license to Springer Nature Switzerland AG 2021
M. Rieger, *Search for $t\bar{t}H$ Production in the $H \rightarrow b\bar{b}$ Decay Channel*, Springer Theses,
https://doi.org/10.1007/978-3-030-65380-4_6

Table 6.1 Run periods, reprocessing versions and run number ranges of the CMS data-taking campaign in 2016. The distinction into eight different periods is mandatory owing to varying detector configurations. Stated luminosities are calculated using Pixel Cluster Counting (PCC) and depend on centrally performed data quality assurance methods as well as the trigger strategy applied in this analysis. Data files are used in the MiniAOD format [4]

Run period	Re-reconstruction version	Run range	$\mathcal{L}$ / fb^{-1}
Run2016B	03Feb2017_ver2-v2	272007–275376	5.75
Run2016C	03Feb2017-v1	275657–276283	2.57
Run2016D	03Feb2017-v1	276315–276811	4.24
Run2016E	03Feb2017-v1	276831–277420	4.03
Run2016F	03Feb2017-v1	277772–278808	3.10
Run2016G	03Feb2017-v1	278820–280385	7.58
Run2016H	03Feb2017_ver{2,3}-v1	280919–284044	8.65
			$\sum = 35.92$

Due to technical stops and varying detector configurations, the data-taking period is divided into eight run periods Run2016B-H, which are summarized in Table 6.1.

Therefore, measurement conditions can vary slightly between different run periods which must be reflected accordingly in the analysis to ensure a reasonable agreement between data and simulation, independent of the run period. For every data sample, the central "re-reconstruction" campaign labeled as 03Feb2017 is used, containing refined detector calibrations based on updated reconstruction software. The integrated luminosities given in Table 6.1 are calculated with the Pixel Cluster Counting (PCC) method [3] that employs "van der Meer" (vdM) scans [5, 6] (cf. Sect. 3.2.5), where only high-quality events recorded with fully operational subdetectors are taken into account.

In general, luminosities depend on the trigger strategy as well as on the choice of primary datasets that are defined by utilizing certain key characteristics of an event such as the distinct presence of one or two isolated leptons. Relevant datasets and corresponding numbers of recorded events are shown in Table 6.2. The trigger strategy employed in this analysis is based on combinations of individual trigger paths that result in a constant integrated luminosity of 35.92 fb^{-1} in all channels (cf. Sect. 6.3.3).

6.2 Simulated Event Samples

This section introduces the simulated samples of events that are used in this analysis to model expected background contributions as well as to describe the $t\bar{t}H$ signal processes according to the SM prediction. First, the event simulation procedure applied within the CMS experiment is briefly explained. Detailed overviews of the generator setup and the set of simulated samples are presented thereafter.

Table 6.2 Collision datasets of the CMS experiment that are relevant for this analysis. The content of each dataset follows a general trigger path with ambiguity resolution in case of overlaps. Thus, datasets represent disjoint sets of events. The numbers of total recorded events are summed over all run ranges and depend on implementations of triggers and reconstruction algorithms as part of the reprocessing campaign 03Feb2017 (Table 6.1)

Channel	Dataset name	Total events/10^6
e	SingleElectron	962.13
μ	SingleMuon	804.01
ee	DoubleEG	492.68
$e\mu$	MuonEG	173.23
$\mu\mu$	DoubleMuon	287.06
		$\sum = 2719.11$

CMS Event Simulation

The simulation of proton-proton collision events is conducted in multiple stages. As explained in Sect. 2.2.1, the interaction of two protons is actually considered as a hard scattering processes of two partons by means of the factorization theorem. The probability for partons to possess a certain momentum fraction with respect to the proton is described by parton distribution functions (PDFs). They depend on the type of the scattering parton and the total momentum transfer Q^2 of the interaction. The implementation of this approach is part of MC event generators [7, 8], which model each aspect of an event by taking into account a comprehensive set of matrix elements up to a certain perturbation order. Procedures involving free parameters such as particle decays and the specification of decay angles are described by appropriately distributed random variables [7, 8].

While the interaction scale is larger than the QCD renormalization scale, $Q^2 > \mu_R^2$, the strong coupling constant is small enough for perturbative calculations to converge. The parton showering approach is employed to describe the evolution of the interaction scale down to a cutoff value by recursively adding $q \to qg$ and $g \to q\bar{q}$ splittings. While perturbative calculations expressed through matrix elements often describe additional hard and large-angle emissions with good accuracy, parton showering exhibits a better efficiency regarding soft and collinear radiations. Conceptually, a phase space overlap between the calculation of the matrix element and the simulation of parton showers might occur, which is resolved by matching algorithms [9, 10]. Examples are the MLM [11] and the FxFx matching scheme [12].

When the renormalization scale is reached, the strong coupling gains relevance and the proceeding event description depends on phenomenological methods to model the hadronization process. An example is the Lund string model [13] as implemented in the PYTHIA event generator [14]. Unstable hadrons are considered to decay thereafter so that only stable particles remain, forming the final-state event content. Furthermore, additional stable particles are added to the final state as a result of three residual contamination effects. Remnants of the colliding protons constitute the underlying event and may interact with particles induced by the hard process, multiple partons can scatter within the same proton-proton interaction, and lastly, multiple protons collide during each bunch crossing and cause numerous pileup interactions owing to the instantaneous luminosity being optimized to increase the overall rate of interesting physics processes. Pileup is modeled by simulating and adding minimum-bias events, where the pileup multiplicity per event is based on the distribution in real data.

Finally, given the set of final-state particles following the event generation procedure, the response of the CMS detector is simulated in two steps [15]. First, particles are propagated through a full simulation of the CMS detector geometry using the GEANT4 (v9.4) toolkit [16, 17]. It models the propagation of particles through matter in a stepwise manner involving all forms of interactions with matter, the production of secondary particles, and the influence of electric and magnetic fields. Second, in the digitization step, the readout of all subdetector components is simulated, which should eventually resemble the detector response for real data. The result is a collection of hits in the tracking and muon systems, energy deposits in electromagnetic and hadronic calorimeters, as well as timing information. Thus, the same reconstruction algorithms can be applied to obtain an accurate description of simulated events with the CMS detector (cf. Sect. 3.3).

Simulated Event Samples

Simulated events are, depending on the physics process, simulated with PYTHIA (v8.200) [14], POWHEG (v1 or v2) [10, 18, 19], or MADGRAPH5_aMC@NLO (v2.2.2) [20]. In all samples, parton showering and hadronization are simulated with PYTHIA (v8.200) [14]. The PDF set NNPDF3.0 [21] is used with $\alpha_s = 0.118$ as recommended by the PDF4LHC group for the second run of the LHC [22, 23]. The masses of the top quark and the Higgs boson are set to 172.5 GeV and 125.0 GeV, respectively. In samples involving top quarks, renormalization and factorization scales are set to the top quark mass, $\mu_r = \mu_F = m_t$, and the CUETP8M2T4 set of PYTHIA tuning parameters for parton showering and modeling of the underlying event is utilized [24, 25]. All other samples use the CUETP8M1 tune [25]. A comparison of all differing parameters can be found in Ref. [26]. Table 6.3 presents a summary of all simulated event samples.

In order to accurately model recorded data distributions, simulated events are normalized to their expected contributions using event weights that involve the cross section of the underlying physics process σ_p, the integrated luminosity $\mathcal{L}$, and a per-event weight produced by the respective event generator program. The normalization weight for an event i is given by

$$\omega_{p,i} = \frac{\sigma_p \cdot \mathcal{L} \cdot \omega_{\text{gen},i}}{\sum_j^{N_{p,\text{gen}}} \omega_{\text{gen},j}}, \tag{6.1}$$

where the denominator sums over all $N_{p,\text{gen}}$ generator event weights. For constant generator weights of one, Eq. 6.1 simplifies to $\omega_{p,i} = \omega_p = \sigma_p \mathcal{L}/N_{p,\text{gen}}$. The calculation of corresponding cross sections is to be conducted for identical phase space

Table 6.3 Simulated signal and background samples as used in this analysis. Corresponding processes are defined in Section 2.2.4, with cross sections in Table 2.4. The label column defines the groups of processes depicted in histograms throughout this thesis. The dilepton mass m_{ll} and the sum of transverse jet momenta H_T are quoted in GeV. Common references for POWHEG can be found in Refs. [10, 18, 19], for PYTHIA in Ref. [14], and for MADGRAPH5_aMC@NLO in Ref. [20]

Dataset/channel/phase space		Generator	Label	Color
$t\bar{t}H$	$H \to b\bar{b}$	POWHEG (v2) [27]	$t\bar{t}H$	■
	$H \nrightarrow b\bar{b}$			
$t\bar{t}$	Inclusive	POWHEG (v2) [28]	$t\bar{t}+b\bar{b}$	■
	SL		$t\bar{t}+b/\bar{b}$	■
	DL		$t\bar{t}+2b$	■
	SL, $t\bar{t}$+hf enriched		$t\bar{t}+c\bar{c}$	■
	DL, $t\bar{t}$+hf enriched		$t\bar{t}+$lf	■
Single $t/\bar{t}$	s-channel, $W \to l\nu$	MADGRAPH5_aMC@NLO (v2.2.2)	Single $t/\bar{t}$	■
	t-channel	POWHEG (v2) [29] + MADSPIN [30]		
	tW-channel	POWHEG (v1) [31]		
W+jets	$70 < H_T \leq 100$	MADGRAPH5_aMC@NLO (v2.2.2) + MLM matching [11]		
$W \to l\nu$	$100 < H_T \leq 200$			
	$200 < H_T \leq 400$			
	$400 < H_T \leq 600$			
	$600 < H_T \leq 800$			
	$800 < H_T \leq 1200$			
	$1200 < H_T \leq 2500$			
	$H_T > 2500$			

(continued)

Table 6.3 (continued)

Dataset/channel/phase space		Generator	Label	Color
Z+jets $Z/\gamma^* \to ll\ 5 < m_{ll} \leq 50$	$70 < H_T \leq 100$	MadGraph5_aMC@NLO (v2.2.2) + MLM matching [11]	V+jets	■
	$100 < H_T \leq 200$			
	$200 < H_T \leq 400$			
	$400 < H_T \leq 600$			
	$H_T > 600$			
Z+jets $Z/\gamma^* \to ll\ m_{ll} > 50$	$70 < H_T \leq 100$	MadGraph5_aMC@NLO (v2.2.2)+ MLM matching [11]		
	$100 < H_T \leq 200$			
	$200 < H_T \leq 400$			
	$400 < H_T \leq 600$			
	$600 < H_T \leq 800$			
	$800 < H_T \leq 1200$			
	$1200 < H_T \leq 2500$			
	$H_T > 2500$			
WW	Inclusive	Pythia(v8.200)	Diboson	■
WZ	Inclusive			
ZZ	Inclusive			
$t\bar{t}W$	$W \to l\nu$	MadGraph5_aMC@NLO (v2.2.2) + MadSpin [30] + FxFx matching [12]	$t\bar{t}+V$	■
	$W \to q\bar{q}'$			
$t\bar{t}Z$	$Z \to ll/\nu\nu$	MadGraph5_aMC@NLO (v2.2.2)		
	$Z \to q\bar{q}$			

conditions as specified for the simulation of events. For example, the production of W and Z/γ^* bosons with additional jets, V+jets, is simulated in different intervals of the scalar sum of transverse momenta H_T and therefore, cross sections must be determined accordingly. A summary of all applied cross sections is shown in Table 2.4. Except for $t\bar{t}$ background samples, phase space regions do not overlap so that event weights follow from Eq. 6.1 and contributions of all processes are added.

For the simulation of $t\bar{t}$ processes, five different samples are generated with significantly overlapping phase spaces. Besides an inclusive sample containing all decay channels of the $t\bar{t}$ system, i.e., single-lepton (SL), dilepton (DL), and full-hadron (FH) events, two additional samples are used, which solely contain either SL or DL events. Furthermore, two "$t\bar{t}$+hf enriched" samples are generated with requirements on generator level to simulate only events with additional heavy-flavor jet emissions, constituting the most important background to $t\bar{t}H$ ($H \rightarrow b\bar{b}$) production. Hence, a considerably large amount of $t\bar{t}$+hf events is obtained while simultaneously lowering the consumption of computing time and storage capacity owing to the reduced generation of $t\bar{t}$+lf events. In turn, the increase of event statistics in various partially overlapping phase space regions must be reflected in the determination of event weights (Eq. 6.1).

The general idea of this phase space extension is that all predictions contained in the inclusively simulated event sample are maintained. This can be achieved by artificially dividing the full phase space into disjoint regions, based on generator-level aspects of the samples to join. In case of the five simulated $t\bar{t}$ samples, full-hadron, single-lepton and dilepton events are distinguished, and the two latter channels are divided further using the definition of the five $t\bar{t}$ subprocesses introduced in Sect. 2.2.4, resulting in a total of eleven regions. Now, relative fractions of the different regions are measured on the inclusive sample and included in the determination of event weights. The fractions for the $t\bar{t}$ sample, simulated using the POWHEG (v2) [10, 18, 19, 28] event generator and parton showering as implemented in PYTHIA (v8.200) [14] with the CUETP8M2T4 tune [24, 25], are shown in Table 6.4. Although these values could be interpreted as branching ratios, they depend, in general, on the choice of generator and are therefore referred to as F in the following. Consequently, the weight for a $t\bar{t}$ event i in a region r can be written as

$$\omega_{r,i} = \frac{F_r \cdot \sigma_{t\bar{t}} \cdot \mathcal{L} \cdot \omega_{\text{gen},i}}{\sum_j^{N_{r,\text{gen}}} \omega_{\text{gen},j}}. \tag{6.2}$$

6.3 Event Selection

The event selection strategy defines the phase space of events to be analyzed and aims to suppress background processes while retaining as many signal events as possible (cf. Sect. 4.2.1). Moreover, its purpose is to ensure that the analysis is conducted in a

Table 6.4 Fractions of $t\bar{t}+X$ processes measured in the inclusive, nominal $t\bar{t}$ sample, simulated using POWHEG (v2) [10, 18, 19, 28], PYTHIA (v8.200) parton showering [14] with the CUETP8M2T4 tune [24, 25], and a top quark mass of 172.5 GeV. Statistical uncertainties are negligible as the number of simulated events is sufficiently large (e.g. ~35000 dilepton $t\bar{t}+b\bar{b}$ events)

$t\bar{t}$ channel	$t\bar{t}+b\bar{b}$ (%)	$t\bar{t}+b/\bar{b}$ (%)	$t\bar{t}+2b$ (%)	$t\bar{t}+c\bar{c}$ (%)	$t\bar{t}+$lf (%)	All (%)
DL	0.05	0.17	0.05	0.78	9.46	10.50
SL	0.21	0.79	0.22	3.87	38.71	43.80
FH	0.24	0.90	0.26	4.70	39.60	45.70

region in which the simulation is found to describe recorded data distributions with reasonable accuracy. The following sections present the selection cuts on global event and physics object quantities in both simulated and recorded event samples in order of their application. They are substantially synchronized with recommendations from Physics Object Groups (POG) and Physics Analysis Groups (PAG) within the CMS experiment. The impact of each individual selection step is visualized at the end of this section.

6.3.1 Vertex Selection and MET Filters

The first two selection steps are sensitive to the quality of recorded data events regarding characteristics of the primary proton-proton collision as well as detector effects. In order to prevent potential biases, these steps are applied likewise to simulated event samples.

First, the quality of the reconstructed, primary interaction vertex is evaluated. It is required to have at least four associated tracks, and maximum longitudinal and transversal distances to the interaction point at the center of the detector of 24 and 2 cm, respectively. If the primary vertex fails any of these selection criteria, the event is discarded.

Second, a set of requirements has to be fulfilled regarding the measurement of missing transverse energy. While large values of MET imply the existence of undetectable objects such as neutrinos or, potentially, particles indicating signs of new physics, these measurements are often induced by undesired detector effects. Possible causes are noise from electromagnetic and hadronic calorimeter components, spurious signals from fake interactions, defective detector material or electronics, energetic particles stemming from residual beam halo, or erroneous boundary effects in object reconstruction algorithms [32]. Therefore, a collection of filter algorithms is employed to detect each individual effect, whereby binary filter decisions are encoded in so-called "MET filters". The list of applied MET filters in data and simulation is shown in Table 6.5. Further details about each particular algorithm are presented in Ref. [32]. An event is selected if none of the employed algorithms reveals unusual effects in the measurement of missing transverse energy.

Table 6.5 Names of MET filters that are part of the selection strategy for data and simulation. Details on filter algorithms are given in Ref. [32]

MET filter name	Applied in data	Applied in simulation
Flag_GoodVertices	✓	✓
Flag_GlobalTightHalo2016Filter	✓	✓
Flag_HBHENoiseFilter	✓	✓
Flag_HBHENoiseIsoFilter	✓	✓
Flag_EcalDeadCellTriggerPrimitiveFilter	✓	✓
Flag_eeBadScFilter	✓	
Flag_BadPFMuonFilter	✓	✓
Flag_BadChargedCandidateFilter	✓	✓

6.3.2 Lepton Selection

This analysis investigates events with final states containing one or two leptons, originating from leptonic W boson decays in case of the $t\bar{t}H$ signal process. It focuses on isolated electrons and muons, and disregards decays involving tau leptons. Type and number of leptons in the final state define a logical separation of the phase space to analyze, which is commonly referred to as channels. Events with a single electron or muon (e, μ) constitute the single-lepton channel (SL), whereas two oppositely charged leptons $(ee, e\mu, \mu\mu)$ represent the dilepton channel (DL).

In both channels, two different lepton definitions are introduced which mainly originate from considerations in the DL channel. Here, the selection strategy serves two partially contrary purposes. On the one hand, the purity of isolated leptons should be reasonably high to reduce fake contributions, which can be achieved through tight cuts on kinematic properties such as pseudorapidity and transverse momentum of the lepton. On the other hand, the selection efficiency should be sufficiently large to maintain an appropriate amount of events for further analysis. In case of two leptons, a common approach is a trade-off between purity and efficiency by requiring a "leading" lepton to pass rather tight selection criteria, while they are partially loosened for the "sub-leading" lepton. This construction, however, must be taken into account likewise in the SL channel, as events must exclusively qualify as one of the two channels to prevent double-counting. Therefore, the set of selection criteria defining the sub-leading lepton in the DL channel serves as a "veto" for additional leptons in the SL channel. This means that if an additional lepton is found that passes all requirements for sub-leading leptons, the event is not classified as a single-lepton event, but it is rather discarded or assigned to the DL channel.

The subsequent paragraphs present the selection criteria for both electrons and muons, as well as additional constraints on lepton pairs in the DL channel.

Electron Selection

The selection of electrons commences with reconstructed, isolated electron hypotheses as provided by the Particle Flow (PF) algorithm (cf. Sect. 3.3.3) [33]. A set of selection criteria is applied as recommended by the CMS EGAMMA object group, corresponding to the tight working point for cut-based electron identification [34, 35].

Transverse momenta and absolute pseudorapidities of leading electrons must fulfill $p_T > 30(20)\,\text{GeV}$ and $|\eta| < 2.1(2.4)$ in the SL (DL) channel, while they are less restrictive with $p_T > 15\,\text{GeV}$ and $|\eta| < 2.4$ for sub-leading electrons. All other selection criteria are identical for both definitions, but depend on the pseudorapidity of the electron's supercluster owing to varying ECAL detector components. Clusters with $|\eta_{SC}| < 1.4442$ are located in the ECAL barrel, and in the ECAL endcaps in case of $|\eta_{SC}| > 1.5660$. Electrons traversing trough the gap between barrel and endcap regions fail the selection.

To distinguish prompt electrons emerging from the primary interaction vertex from those originating from hadron decays within jets, the relative isolation variable r_{iso} is constructed. It compares the transverse momentum of the electron to that of charged hadrons (CH), neutral hadrons (NH), and photons located in a cone around the electron with a radius in the η-ϕ-plane of $\Delta R = 0.3$, and is defined as

$$r_{iso} = \frac{p_T^{CH} + \max(E_T^{NH} + E_T^{\gamma} - \rho A,\ 0)}{p_T}. \tag{6.3}$$

The term ρA in the numerator is subtracted from the sums of neutral particles to account for contributions due to pileup. It consists of the energy density ρ measured in the event, and an effective area A, which denotes a calibration measure estimated from simulation. The relative isolation must be smaller than $r_{iso} < 0.0588$ in the barrel, and $r_{iso} < 0.0571$ in the endcap region. In addition, a requirement is posed on the impact parameter of the track associated to the electron, i.e., its closest proximity to the interaction point in longitudinal and transversal direction, d_z and d_R. Electrons are selected if $d_z < 1.0\,\text{mm}$ and $d_R < 0.5\,\text{mm}$ in the barrel, and $d_z < 2.0\,\text{mm}$ and $d_R < 1.0\,\text{mm}$ in the endcap.

Muon Selection

The muon selection starts from hypotheses of global muons as reconstructed by the PF algorithm (cf. Sect. 3.3.3) [33]. The CMS MUON object group recommends a set of cuts on variables of muon tracks measured in the inner tracking system of the detector (cf. Sect. 3.2.1), as well as in the outer muon system (cf. Sect. 3.2.3) [36]. This analyses uses the tight working point, which is documented in Ref. [37].

A calibration of muon momenta is performed via the so-called "Rochester" correction method. It aims to compensate for biases in the momentum measurement caused by misalignment of detector components, uncertainties in the magnetic field, or biases in reconstruction algorithms, and is applied to both data and simulated events. First, an idealized sample of $Z/\gamma^* \to \mu^+\mu^-$ events is simulated and transverse muon momenta on generator-level are smeared using a functional relation that resembles the experimental resolution as a function of pseudorapidity. The resulting sample represents a perfectly aligned detector. Then, by comparing these distributions between data and actual simulation in granular bins of charge, pseudorapidity, and azimuthal angle, momentum scale correction factors can be extracted. In the second step, the corrections are fine-tuned so that the mass of the lepton pair is stable in all employed bins to account for variations of the muon detection efficiency. A detailed description of the method is given in Ref. [38]. The correction parameters are determined and validated in a central effort of the CMS MUON object group.

Cuts on the transverse momentum and the absolute pseudorapidity differ between the leading and sub-leading muon definition. A successful selection requires leading muons to fulfill $p_T > 26(25)$ GeV and $|\eta| < 2.1(2.4)$ in the SL (DL) channel, whereas sub-leading muons must fulfill $p_T > 15$ GeV and $|\eta| < 2.4$.

Similar to electrons, a relative isolation variable is constructed to identify prompt muons while suppressing muons from weak hadron decays within jets [39]. It compares the transverse muon momentum p_T to contributions from charged (CH) and neutral hadrons (NH), as well as photons in a cone around the muon momentum direction with radius of $\Delta R = 0.4$, and is defined as

$$r_{\text{iso}} = \frac{p_T^{\text{CH}} + \max(E_T^{\text{NH}} + E_T^{\gamma} - p_T^{\text{PU}}/2, \, 0)}{p_T}. \tag{6.4}$$

The subtracted term $p_T^{\text{PU}}/2$ describes a pileup correction to the sums of energies of neutral particles. It is estimated from charged, hadronic calorimeter deposits, which can be associated to pileup vertices [39]. The factor $1/2$ is extracted from simulation and denotes the multiplicity ratio between produced neutral particles and charged hadrons in inelastic proton-proton collisions [39]. Leading muons in the SL (DL) channel are accepted if their relative isolation is smaller than $r_{\text{iso}} < 0.15(0.25)$, whereas sub-leading muons must fulfill $r_{\text{iso}} < 0.25$.

Additional Criteria in the Dilepton Channel

Except for s- and t-channel single $t/\bar{t}$ production, events with two isolated leptons can potentially emerge from almost all processes considered in this analysis. However, further selection criteria can cause a significant reduction of some background contributions. An evident requirement is that the two leptons must possess opposite electrical charges. This way, approximately 50% of events with a single, genuine lepton and an additional fake lepton are suppressed.

The invariant mass m_{ll} of the dilepton system provides another suitable handle. Events from heavy-flavor resonances as well as Drell-Yan processes with low masses, i.e., photons decaying into pairs of leptons, are suppressed by requiring $m_{ll} > 20$ GeV. Furthermore, high-mass Drell-Yan processes are rejected by excluding a mass window around the Z boson mass of 76 GeV $< m_{ll} < 106$ GeV for events with same-flavored leptons (ee, $\mu\mu$). Effectively, these criteria suppress contributions from Z+jets and diboson processes.

Table 6.6 Summary of applied high-level trigger paths and corresponding integrated luminosities. For channels with multiple paths, the trigger rule is constructed using OR concatenations

Trigger channel	Trigger paths	$\mathcal{L}/\text{fb}^{-1}$
e	HLT_Ele27_WPTight_Gsf	35.92
μ	HLT_IsoMu24	35.92
	HLT_IsoTkMu24	
ee	HLT_Ele23_Ele12_CaloIdL_TrackIdL_IsoVL_DZ	35.92
$e\mu$	HLT_Mu23_TrkIsoVVL_Ele12_CaloIdL_TrackIdL_IsoVL{_DZ}	35.92
	HLT_Mu8_TrkIsoVVL_Ele23_CaloIdL_TrackIdL_IsoVL{_DZ}	
$\mu\mu$	HLT_Mu17_TrkIsoVVL_Mu8_TrkIsoVVL{_DZ}	35.92
	HLT_Mu17_TrkIsoVVL_TkMu8_TrkIsoVVL{_DZ}	

6.3.3   Trigger Selection

Simulated and recorded events are required to pass an online selection procedure performed by the two-tiered trigger system of the CMS detector (cf. Sect. 3.2.4). Different trigger configurations are encoded in so-called "trigger paths" and individual selection or rejection decisions are stored per event. The list of required paths is summarized in Table 6.6.

Events must pass at least one trigger path. The distinct association to a particular analysis channel is premised upon further criteria as part of the lepton selection (cf. Sect. 6.3.2), which generally involves stricter selection cuts to prevent turn-on effects with respect to online trigger requirements. The logical OR concatenations of trigger paths per channel were fully operational and not pre-scaled during the data-taking period in 2016. Therefore, as presented in Table 6.2, the integrated luminosity amounts to $35.92\,\text{fb}^{-1}$ and is coherent in each channel.

Events in the e and μ channels are selected using single-lepton online triggers that require exactly one electron or muon with a transverse momentum of at least 27 GeV or 24 GeV, respectively. Dilepton triggers are used to record events in the ee, $e\mu$, and $\mu\mu$ channels. In same-flavor channels, the leading electron (muon) should exhibit a transverse momentum of at least 23 (17) GeV, and at least 12 (8) GeV in case of the sub-leading electron (muon). The trigger in the $e\mu$ channel requires either a muon with at least 23 GeV and an electron with at least 12 GeV, or a muon with at least 8 GeV and an electron with 23 GeV or greater.

In general, an event might be accepted by multiple triggers, for instance when an event contains an energetic muon that passes the single muon trigger criteria, and an additional muon with a transverse momentum between the lower threshold of the dimuon trigger, and the threshold of the single-muon trigger. This ambiguity is resolved by performing the offline lepton selection (cf. Sect. 6.3.2) and comparing the resulting channel to that implied by the trigger decisions, labeled "Trigger channel" in the left column of Table 6.6.

Typically, lepton and trigger channels are required to be identical, and, in case of real data events, need to match the primary dataset as presented in Table 6.2. For recorded dilepton events, however, a different approach is pursued in order to increase the event statistics by mitigating potential dilepton trigger inefficiencies. Data events contained in the `SingleElectron` (`SingleMuon`) dataset (Table 6.2), which are identified as ee or $e\mu$ ($\mu\mu$ or $e\mu$) events by means of the lepton selection, are retained in case they do not pass single-electron (single-muon) trigger requirements. This trigger strategy was developed alongside measurements of inclusive and differential $t\bar{t}$ cross sections with the CMS experiment [40–42]. As a result, the number of events increases by 9.5% in the ee channel, by 9.3% in the $e\mu$, and by 6.6% in the $\mu\mu$ channel.

6.3.4 Missing Transverse Energy Selection

As explained in the discussion of the event signatures of the $t\bar{t}H$ process and potential background processes in Sect. 2.2.4, either one (SL channel) or two neutrinos (DL channel) are produced in the two W boson decays. These neutrinos interact only very weakly and leave the experiment undetected. Since the initial state of the hard parton scattering process is approximately balanced in the transverse plane, an imbalance in the measurement of transverse momenta of all reconstructed objects provides a handle on activity of the neutrinos in an event (cf. Sect. 3.3.5).

In order to suppress background contributions that are not induced by processes involving W bosons originating from top quark decays, a cut on the amount of missing transverse energy is applied. Therefore, especially multijet background from QCD interactions is heavily suppressed, which is difficult to simulate with an adequate amount of statistics due to its high cross section and vanishingly low selection efficiency. Single-lepton events are required to exhibit a missing transverse energy of at least 20 GeV. Dilepton events with same-flavored leptons, i.e., ee and $\mu\mu$, must contain at least 40 GeV of missing transverse energy. The cut is omitted for $e\mu$ events with opposite lepton flavors as no background process without a W boson pair exists that could produce exactly one isolated electron and muon with opposite charges.

6.3.5 Jet and b-Tag Selection

When considering the nominal final state of the $t\bar{t}H$ signal process, one expects six jets in the single-lepton, and four jets in the dilepton channel, with four of them originating from decays of b hadrons. With the exception of contributions from $t\bar{t}+b\bar{b}$, the remaining background processes are likely to exhibit fewer jets or b tags and can therefore be suppressed by appropriate selection cuts on jet and b tag multiplicities. The accurate reconstruction of jets and the efficient identification of b-flavored jets is crucial for the sensitivity of analyses involving a high number of jets. Furthermore, the jet multiplicity constitutes the basis for the categorization of events as described in Sect. 4.2.2.

Jet selection criteria and the method for calibrating jet energy resolution are discussed in the next paragraph, followed by the requirements for performing b-tagging on selected jets.

Jets Selection

Jets are reconstructed from particles that are not identified by the PF algorithm as isolated leptons, prompt photons, or particles from pileup events by means of charged hadron subtraction (cf. Sect. 3.3.4) [33]. To construct a jet hypothesis, these particles are clustered using the IRC-safe anti-k_T algorithm with a distance parameter $\Delta R = \sqrt{\Delta \eta^2 + \Delta \phi^2} = 0.4$ [33, 43, 44], implemented in the FASTJET package [45]. A set of basic selection criteria is applied that complies with recommendations of the CMS JETMET object group at the loose working point [46]. It requires that the jet is clustered from two or more constituents of which at least one is charged, energy deposits in the hadronic calorimeter that are linked to inner tracks, and that energy fractions due to neutral hadronic and electromagnetic activity are below 99%.

The jet energy scale in data and simulated events is calibrated in a manifold calibration procedure [44, 47, 48] (cf. Sect. 3.3.4). The jet energy and momentum resolution is found to be lower in data than in simulated events. Therefore, a hybrid resolution smearing method is employed to compensate for this effect in simulations by scaling the reconstructed four-momentum of a jet with a scaling factor f [44, 47, 48]. First, the representation of a jet on generator-level is determined through a matching that obeys $\Delta R(\text{jet}, \text{jet}^{\text{gen}}) < 0.2$ and $|p_T - p_T^{\text{gen}}| < 3\sigma p_T$, where σ denotes the relative momentum resolution as measured in data. Then, in case a generator jet is found, the scaling factor is constructed from the functional relation [44]

$$f = 1 + (c - 1)\frac{p_T - p_T^{\text{gen}}}{p_T} \tag{6.5}$$

with c being a calibration factor of the method. Otherwise, f results from a random smearing approach [44],

$$f = 1 + N(0, \sigma)\sqrt{\max(c^2 - 1, \, 0)}, \tag{6.6}$$

where N is sampled from a normal distribution with a mean of zero and a variance of σ^2.

Following the jet energy scale and momentum resolution calibrations, selection criteria are applied on kinematic variables of a jet, as well as on the number of jets in an event. In the single-lepton channel, events must contain at least four jets with $p_T > 30\,\text{GeV}$ and $|\eta| < 2.4$. Moreover, a jet is rejected if it is located within a radius of $\Delta R = 0.4$ around a selected, isolated lepton as discussed above (cf. Sect. 6.3.2). In the dilepton channel, an event is accepted if it contains at least four jets with $p_T > 20\,\text{GeV}$ and $|\eta| < 2.4$, with the two most energetic jets also fulfilling $p_T > 30\,\text{GeV}$. Similar to the rationale concerning the two differing lepton definitions, this distinction between leading and sub-leading jets is introduced in order to achieve a reasonable trade-off between event purity and selection efficiency in the dilepton channel.

Identification of b-Flavored Jets

Jet flavor algorithms are used to identify jets from decays of long-lived b hadrons emerging from bottom quarks in the hard scattering process. They travel a significant distance with respect to the primary interaction vertex before eventually decaying, typically ~ 1 cm at a momentum of ~ 100 GeV, which is resolvable by means of secondary vertex reconstruction algorithms (cf. Sect. 2.46).

In this analysis, b jets are identified with the second revision of the Combined Secondary Vertex (CSVv2) algorithm [49, 50]. Selected jets are attributed a b tag if their assigned CSV discriminant is larger than or equal to 0.8484. This criterion corresponds to the medium working point, defined by a misidentification probability of 1% as measured on a simulated $t\bar{t}$ sample [50]. The associated tagging efficiency was found to be $\sim 64\%$ [50].

In all channels, events are rejected if they contain less than three selected, b-tagged jets. In the dilepton channel, at least one b tag must be attributed to a leading jet. Considering the nominal tagging and misidentification efficiencies, the combined b-tagging selection efficiency amounts to 27.5% (55.2%) for events with three (four) true b-flavored jets (Eq. 4.1). This way, background processes that nominally entail a lower b tag multiplicity, such as $t\bar{t}$+lf, single $t/\bar{t}$, V+jets, and especially multijet events from QCD interactions, are significantly suppressed.

6.3.6 Selection Results

After the application of all selection criteria as discussed in the previous sections, a certain number of generated events remain for further analysis. They are presented in Table 6.7 for the $t\bar{t}H$ signal process, each individual background process, and recorded data. Furthermore, they are divided according to the first categorization step utilizing the number of jets in either the single-lepton or dilepton channel (cf. Sect. 4.2.2). Most notably, the number of remaining signal and $t\bar{t}$ background events is sufficient for the training of deep neural networks, which are employed for the second event categorization step on the basis of a predicted, most probable process (cf. Chap. 7). The relative normalization of simulated events for accurately modeling recorded data distributions is presented in the following sections after a discussion of simulation corrections.

The impact of each individual selection step is visualized in Fig. 6.1, separately for each channel. In the two single-lepton channels, less than 2% of the events are discarded due to MET filtering and vertex selection. The largest influence results from the lepton selection after which only ~ 8 (10)% of signal (background) events remain in the e channel, and only ~ 11 (14)% in the μ channel. The differences stem from a slightly better identification and reconstruction efficiency for muons (cf. Sects. 3.3.3 and 3.3.2). The same argument holds for the trigger selection, which has

Table 6.7 Numbers of generated and recorded events that pass all selection criteria, evaluated separately for the four jet categories in the single-lepton and dilepton channels

Process	Single-lepton			Dilepton
	$4j$	$5j$	$\geq 6j$	$\geq 4j$
$t\bar{t}+b\bar{b}$	28230	46989	79737	20772
$t\bar{t}+b/\bar{b}$	65167	64711	56493	21274
$t\bar{t}+2b$	28754	31818	35462	9654
$t\bar{t}+c\bar{c}$	56498	58002	60797	23294
$t\bar{t}+\mathrm{lf}$	276888	157240	92248	27173
Single $t/\bar{t}$	1246	677	453	47
V+jets	5834	6102	9207	657
Diboson	41	16	6	2
$t\bar{t}+V$	18470	28733	66760	18458
$t\bar{t}H$	35274	59834	99280	23442
Data	22900	17097	14057	3332

a rather small impact when applied after the lepton selection. Up to the selection of missing transverse energy, the overall efficiency is slightly larger for background processes. After selecting jets and b tags, however, the number of background events in the e (μ) channel is reduced by another 89% (88%), while only 51% (51%) of signal events are discarded.

With the exception of a few aspects, a similar reasoning applies to the three dilepton channels. In the same-flavor channels (ee, $\mu\mu$), the lepton selection exhibits a significant difference between signal and background processes. This effect is caused by Z+jets events with $Z/\gamma^* \rightarrow e^+e^-/\mu^+\mu^-$ that yield higher selection efficiencies owing to a less populated final state. However, their contribution is eventually reduced after applying the Z boson mass veto. After the last selection stage, the signal efficiency exceeds that of background events in relative terms by 39% for ee, by 34% for $e\mu$, and by 40% in $\mu\mu$ events.

6.4   Corrections to Simulated Events

This section presents a series of corrections that are applied to simulated events in order to resolve discrepancies observed in the comparison to data. Potentially, these discrepancies could be caused by various effects within the event simulation and reconstruction procedures, starting from MC event generator tunings, to parton showering, hadronization, over to the detector simulation and reconstruction algorithms.

In the scope of proton-proton collision analyses, a typical approach for deriving corrections is the comparison of characteristic quantities between data and simulation

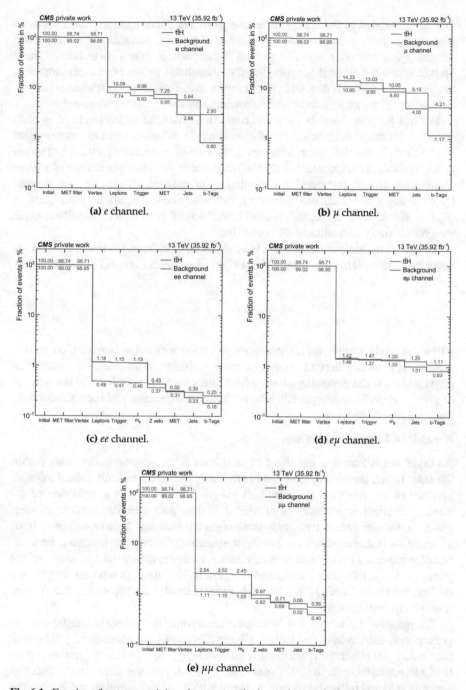

Fig. 6.1 Fraction of events remaining after consecutively passing nominal selection steps, evaluated for the $t\bar{t}H$ signal process (red) and all background processes combined (blue), which is dominated by $t\bar{t}$ owing to the large amount of simulated events. It should be noted that selection steps differ between channels. After the last step ("b-Tags"), the fraction of remaining events is larger for signal than for background events in all channels

that are correlated to the effects to be compensated. More precisely, disagreement observed in these quantities does not represent a cause, but rather a consequence of underlying mismodeling. The general assumption is that a correction of these derived quantities through a reweighting of simulated events effectively improves the agreement between data and simulation in distributions of correlated variables. Consequently, a thorough treatment of uncertainties is absolutely necessary.

For this purpose, identification and reconstruction efficiencies are used in most cases, which can be extracted in simulation and data. Whereas generator information is exploited for simulations, efficiencies in data are often measured through tag-and-probe methods. Disagreements between efficiencies, for example in case of trigger efficiencies, directly translate into deviating event rates, which must be compensated by including derived data-to-simulation ratios to event weights as introduced in Eq. 6.1. The term "scale factor" is used interchangeably to denote additional event weights resulting from simulation corrections.

In case of multiple corrections, it is advantageous to factorize the impact on the nominal event weight $\omega_{\sigma \mathcal{L}}$. Hence, the weight of an event i (Eq. 6.1) is amended to

$$\omega_i = \omega_{i,\sigma \mathcal{L}} \cdot \prod_c^{\text{corrs.}} \omega_{i,c}, \qquad (6.7)$$

where the product runs over all incorporated corrections c. Furthermore, correction weights $\omega_{i,c}$ usually depend on one or more variables of an event or a contained physics object. The following paragraphs address corrections related to the number of pileup interactions, trigger efficiencies, lepton efficiencies, and the calibration of the b-tagging discriminant.

Number of Pileup Interactions

As explained in Sect. 6.2, the effect of additional proton-proton interactions within the same bunch crossing is modeled in simulated events by including the final-state particles of a number of minimum-bias events. The probability distribution of the number of pileup interactions is measured in data with a limited amount of integrated luminosity before simulation campaigns are initiated. Volatile run conditions affecting the instantaneously luminosity, the number of circulating bunches, the number of protons per bunch, and further beam parameters can cause variations of the pileup profile in data over a measurement period. Therefore, correction weights are derived and applied to simulated events in order to resolve discrepancies in the pileup multiplicity distribution.

The number of pileup events in simulation depends on the total inelastic proton-proton cross section at $\sqrt{s} = 13\,\text{TeV}$. Figure 6.2a shows the normalized pileup profiles in data and simulation for a cross section of $\sigma_{\text{inel.}} = 69.2\,\text{mb}$ as recommended by the CMS experiment for the 2016 data-taking period. The ratio between the data and simulation yields pileup correction weights, which are visualized by the blue curve in Fig. 6.2b and exhibit values close to one for the majority of events. The uncertainty on the inelastic cross section of $\pm 4.6\%$ [51] is propagated to the weight distribution, denoted by the green and red curves. After the realization of this analysis, the

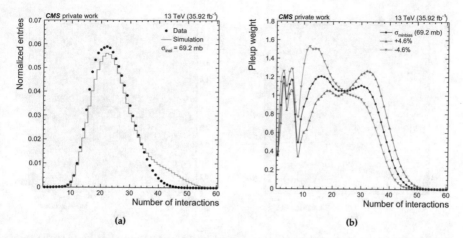

Fig. 6.2 Distribution of the number of pileup interactions per bunch crossing in simulation and data, assuming an inelastic proton-proton cross section of 69.2 mb (**a**), and the resulting pileup weight, i.e., the ratio between measured and simulated interactions (**b**). The systematic uncertainty on the pileup weight is computed by varying the cross section by ±4.6% [51]

CMS experiment published a new measurement of the inelastic proton-proton cross section, stating a value of 68.6 ± 0.5 (syst) ± 1.6 (lumi) mb [52]. The difference to the utilized cross section is marginal and well covered by experimental uncertainties due to systematic effects and uncertainties of the luminosity determination.

Trigger Efficiencies

Efficiencies of the single-lepton and dilepton triggers applied in the selection strategy of this analysis (cf. Sect. 6.3.3) were measured by the CMS EGAMMA [35], MUON [53], and TOP groups [40–42]. The comparison between efficiencies in data and simulation yields slight discrepancies, which can be compensated by the introduction of event weights, so-called trigger scale factors. It should be noted that all electron and muon selection criteria outlined in Sect. 6.3.2 are identical to those employed in the efficiency measurements listed above. Therefore, derived weights are applicable to the phase space of the presented analysis without the need for extrapolation measures.

In the single-lepton channel, scale factors are determined using tag-and-probe methods in $Z/\gamma^* \to e^+e^-$ or $Z/\gamma^* \to \mu^+\mu^-$ events. In order to obtain pure data samples, selection criteria are adapted to characteristics of Z boson decays, such as opposite lepton charges and a mass m_{ll} of the dilepton pair close to the Z boson mass. Contamination uncertainties are estimated by varying these criteria. To measure the trigger efficiency, "trigger elements", i.e., individual online leptons that would cause the trigger to accept the event, are built and matched to leptons resulting from the offline selection using a ΔR criterion. Events are considered if they contain at least one matched "tag" lepton. The efficiency follows from the measurable frequency of events with an additionally matched "probe" lepton.

Resulting scale factors between data and simulation are shown in Fig. 6.3 for the single-electron trigger (left) and the single-muon trigger (right). For electrons, values

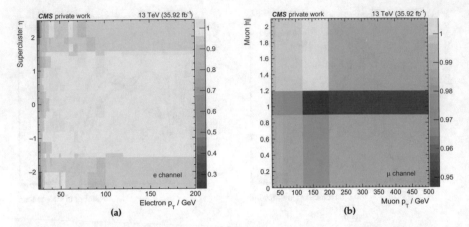

Fig. 6.3 Trigger scale factors as used in the e channel (**a**) [35] and in the μ channel (**b**) [53]. Values in the e channel depend on p_T and η_{SC} of the supercluster of the electron. They are almost flat and close to one with minor reduction for high η_{SC}, as well as for low p_T due to possible turn-on effects. Values in the μ channel depend on p_T and $|\eta|$ of the muon. They are also close to one with small drops at $|\eta| \approx 1$, caused by the barrel-endcap transition in the muon system (cf. Sect. 3.2.3). A slight turn-on effect is visible in the low p_T regime

are parametrized by the transverse momentum p_T and the pseudorapidity η_{SC} of the assigned supercluster. They are almost flat in both variables and close to one, except for the low p_T regime due to possible turn-on effects at the trigger threshold. For muons, scale factors are parametrized by the transverse momentum and absolute pseudorapidity, with values fairly close to one and a minor turn-on effect in the low p_T regime. At $|\eta| \approx 1$, it drops to $\sim$95% which is caused by the transition between barrel and endcap regions in the muon system (cf. Sect. 3.2.3). In both cases, scale factors for leptons that exceed the respective upper p_T bound are extracted from the right-most bin at the given pseudorapidity.

In the dilepton channel, scale factors are measured by monitoring the rates of "cross triggers", i.e., trigger paths that are weakly correlated to the applied dilepton triggers. This study was performed within analyses investigating total and differential $t\bar{t}$ cross sections with the CMS experiment in a compatible phase space [40–42]. First, the number of events $N_{X,t\bar{t}}$ that are accepted by the cross trigger and pass the $t\bar{t}$ selection criteria is determined. Then, the fraction of events that additionally pass the respective dilepton trigger,

$$\epsilon_{DL} = \frac{N_{X+DL,t\bar{t}}}{N_{X,t\bar{t}}}, \tag{6.8}$$

yields a robust estimate for the efficiency of the dilepton trigger. The main challenge arises from the selection of appropriate cross triggers that exhibit a sufficient amount of statistics. The cited efficiency measurement uses a combination of five trigger

Fig. 6.4 Trigger scale factors as used in the dilepton channels and measured in the scope of analyses investigating total and differential $t\bar{t}$ cross sections with the CMS experiment in a compatible phase space [40–42]. Their values depend on the absolute pseudorapidities of the leading ($|\eta_1|$) and sub-leading lepton ($|\eta_2|$), respectively, where the bin range $|\eta| < 2.4$ implies $|\eta| \geq 1.2$. Central values are close to one with uncertainties mostly below 1%

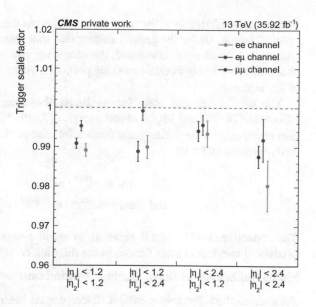

paths that are sensitive to missing transverse energy in the event. Uncertainties are conservatively estimated using Clopper–Pearson intervals [54].

Resulting scale factors between data and simulation efficiencies are presented in Fig. 6.4. In all lepton channels (ee, $e\mu$, $\mu\mu$), scale factors are parametrized by the absolute pseudorapidity of the lepton with the higher transverse momentum $|\eta_1|$, and that of the other lepton $|\eta_2|$. Both axes are divided into only two bins, ranging from $0 \leq |\eta| < 1.2$ and $1.2 \leq |\eta| < 2.4$. Values are consistently between $\sim$98% and $\sim$100% with uncertainties mostly below 1%, implying only slight discrepancies related to the trigger selection in the dilepton channel.

Lepton Efficiencies

Similar to the examination of lepton-dependent trigger efficiencies, the rates of leptons passing specific stages in reconstruction and selection procedures are monitored and compared between data and simulation. By using similar tag-and-probe techniques, corresponding efficiencies can be computed to construct ratios that yield parametrized scale factors. The procedures are typically factorized and divided into separate efficiency measurements to achieve a more granular control over possible discrepancies and resulting corrections.

For electrons, the measurement consists of the determination of the reconstruction efficiency, and of the identification and selection efficiency [35]. Muon efficiencies are separately measured for tracking criteria [55], reconstruction and identification procedures, and isolation requirements [37]. In all cases, tag-and-probe methods are deployed using $Z/\gamma^* \rightarrow e^+e^-/\mu^+\mu^-$ events. Purity in data samples is ensured through requirements on opposite lepton charges and a dilepton mass m_{ll} close to the mass of the Z boson. One lepton is required to pass tight criteria of the respective method. It is referred to as the "tag" lepton, whereas the other lepton constitutes the "probe" lepton. Under the general assumption that selection decisions between tag and probe leptons are uncorrelated, the efficiency can be inferred from the ratio of probe leptons passing a certain working point of a method with respect to the number of tag leptons.

The efficiencies and scale factors listed above were measured by the CMS EGAMMA [34, 35] and MUON object groups [37, 39, 55] for identical lepton selection requirements. Resulting scale factors for electrons and muons are obtained by multiplication and read

$$\omega_e = \omega_e^{\text{reco}} \cdot \omega_e^{\text{ID}} \tag{6.9}$$

$$\text{and} \quad \omega_\mu = \omega_\mu^{\text{track}} \cdot \omega_\mu^{\text{reco+ID}} \cdot \omega_\mu^{\text{iso}}. \tag{6.10}$$

The corrective event weight is either ω_e or ω_μ in events with a single lepton, or the product of two lepton scale factors in the dilepton channel.

Shape Calibration of the b-Tagging Discriminant

Another method that might exhibit discrepancies between data and simulation is the CSVv2 algorithm used for the identification of b-flavored jets (cf. Sect. 3.3.6) as the underlying machine learning approach incorporates numerous input variables that could potentially be subject to mismodeling [50]. The output of the algorithm, referred to as the CSV discriminant, can be used in two different ways. Firstly, a specific threshold value can serve as a binary criterion for the definition of a b tag, entailing specific identification and misidentification efficiencies. Secondly, the shape information of the CSV discriminant can be exploited further. This analysis utilizes binary b jet identification and CSV discriminant values as potential inputs to the neural network techniques for event categorization and process separation (cf. Chap. 7). Therefore, it is not sufficient to infer event weights that solely compensate for deviating b-tagging efficiencies, but a calibration of the CSV discriminant shape is necessary.

The shape calibration is performed by means of an iterative fitting procedure [50], which was first described in Ref. [56]. It aims to extract a scale factor f per jet that depends on its flavor, transverse momentum, pseudorapidity, and, most notably, the CSV discriminant itself. The true jet flavor is determined using hadron flavor information following the "ghost association" method [50, 57]. Equivalently to the factorized lepton scale factors in the dilepton channel, the corrective event weight ω_{CSV} is constructed from the product of scale factors of all selected jets in an event,

$$\omega_{\text{CSV}} = \prod_j^{\text{jets}} f_j, \tag{6.11}$$

and is eventually applied to simulated events. It should be noted that the method is not supposed to change the expected rate of events as determined prior to the application of any b tag selection criteria, which potentially requires an a-posteriori reweighting of events in bins of jet multiplicity.

The iterative calculation of jet scale factors f proceeds as follows. Two regions are defined that are either enriched with b jets, or jets induced by gluons or u, d, or s quarks ($udsg$ jets). No dedicated region is included for jets originating from c quarks. The former region, called heavy-flavor (HF) region, is obtained from a selection strategy that favors dilepton events from $t\bar{t}$ processes. To reduce Z+jets background contributions in channels with same-flavored leptons, events are discarded in case the dilepton mass m_{ll} is within 10 GeV around the Z boson mass, or the missing transverse energy is below 30 GeV. The resulting event sample is composed of $t\bar{t}$ ($\sim$87%), single $t/\bar{t}$ ($\sim$6%), and Z+jets ($\sim$7%) events [50]. For the light-flavor (LF) region, Z+jets events are selected with either two oppositely charged electrons or muons in the final state. Selection criteria on the dilepton mass and missing transverse energy are inverted with respect to the HF region, resulting in a very pure Z+jets sample ($\sim$99.9%) [50].

In both regions, events are required to contain exactly two jets to enable tag-and-probe methods. The tag jet in the HF region is required to pass the medium working point of the CSV algorithm in order to consider the remaining jet for probing. Vice versa, the tag jet in the LF region is to be vetoed by means of the loose working point. As a result, the CSV discriminant of the probe jet can be examined for the extraction of scale factors in bins of its p_T, η, and the discriminant value. Owing to the fact that the correction is only supposed to perform a shape calibration rather than a rate correction, expected yields in simulated events are normalized to data across all CSV discriminant bins within the same p_T and η intervals.

Driven by the Z+jets contamination in the HF region, an interdependence of the scale factors is introduced. Before data-to-simulation ratios can be deduced, the respective "cross" contamination is subtracted from the measured number of events in data. The contamination itself, however, depends on the weights ω_{CSV} of events in the opposite region. This interdependence is resolved through an iterative definition of jet scale factors f_i at a certain iteration step i as

$$\text{(HF region)} \qquad f_{b,i} = \frac{\text{Data} - \vec{\omega}_{i-1} \cdot \vec{\text{MC}}_{udsg+c}}{\text{MC}_b} \qquad (6.12)$$

$$\text{(LF region)} \qquad f_{udsg,i} = \frac{\text{Data} - \vec{\omega}_{i-1} \cdot \vec{\text{MC}}_{b+c}}{\text{MC}_{udsg}} \qquad (6.13)$$

in a particular measurement bin. Contaminations from c jets are subtracted universally. The notation $\vec{\omega}_{i-1} \cdot \vec{\text{MC}}$ signifies that each simulated MC event is multiplied with its individual weight of the previous iteration, initialized with one, whereas remaining terms denote scalar event counts. In fact, the product is identical to the sum of weights of relevant contamination events, including normalization terms as explained above. After $\sim$3 iterations, scale factors appear to converge. For $udsg$ jets, a fit of a sixth-degree polynomial to the scale factor profile, excluding the first and last bins, is used to extract a functional relation between the value of the CSV discriminant and the jet scale factor. No satisfying parameterization was found for b

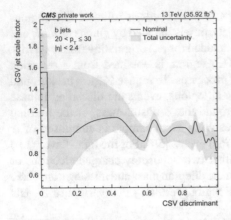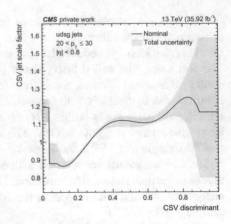

Fig. 6.5 Jet scale factors resulting from the CSV discriminant shape calibration for b jets with $20 < p_T/\text{GeV} \leq 30$ and $|\eta| < 2.4$ (left), and for $udsg$ jets with $20 < p_T/\text{GeV} \leq 30$ and $|\eta| < 0.8$ (right). Gray areas represent total uncertainties of the scale factor calculation. The flavor is determined using hadron flavor information following the "ghost association" method [50, 57]

jets so that a simple smoothing between bin contents is employed instead. Figure 6.5 shows exemplary scale factors for b jets (left) and for $udsg$ jets (right).

Eventually, corrective event weights are computed using Eq. 6.11. In order to validate their impact in a separate phase space, weights were applied to single-lepton $t\bar{t}$ events. The selection required events to contain exactly four jets of which two are b-tagged according to a medium working point. The agreement of the CSV discriminant between data and simulation was found to be improved.

Uncertainties inherent to the calibration procedure comprise variations due to jet energy scales, sample purities, as well as statistical uncertainties resulting from limited numbers of events in each measurement bin. Additionally, uncertainties for c jets are estimated conservatively by assigning twice the relative uncertainty as obtained for b jets, adding in quadrature all uncertainties in the same bin, while setting their nominal scale factors to one. A detailed description of all uncertainties is provided in Ref. [50].

6.5   Evaluation of Background Modeling

After applying event corrections due to pileup interactions, trigger and lepton efficiencies, and the calibration of the b-tagging discriminant shape in form of multiplicative, corrective event weights, simulated events can be normalized to their expected contributions for an integrated luminosity of $\mathcal{L} = 35.92\,\text{fb}^{-1}$ (Eq. 6.7).

Table 6.8 shows a comparison of expected yields separately for the four jet categories in the single-lepton and dilepton channels. Stated uncertainties result from the limited statistics of simulated events, prior to the fit to data. In total, 58893 ± 174

Table 6.8 Observed and expected event yields prior to the fit to data, and quoted separately for the four jet categories in the single-lepton and dilepton channels. Expected yields result from the total numbers of simulated event passing the event selection (Table 6.7) and by applying the product of all event weights (Eq. 6.7). Stated uncertainties denote only statistical uncertainties due to the limited number of simulated events

Process	Single-lepton		Dilepton	
	$4j$	$5j$	$\geq 6j$	$\geq 4j$
$t\bar{t}+b\bar{b}$	837 ± 7	1422 ± 10	2454 ± 13	572 ± 6
$t\bar{t}+b/\bar{b}$	1941 ± 11	1936 ± 11	1722 ± 11	575 ± 6
$t\bar{t}+2b$	842 ± 7	946 ± 8	1050 ± 8	259 ± 4
$t\bar{t}+c\bar{c}$	2916 ± 19	3146 ± 20	3360 ± 21	734 ± 7
$t\bar{t}+$lf	15784 ± 45	8996 ± 35	5323 ± 27	855 ± 8
Single $t/\bar{t}$	1007 ± 33	590 ± 26	407 ± 22	51 ± 8
$V+$jets	355 ± 10	224 ± 7	169 ± 5	14 ± 2
Diboson	17 ± 3	8 ± 2	2.3 ± 0.9	1.2 ± 0.8
$t\bar{t}+V$	62 ± 2	92 ± 3	184 ± 4	40 ± 2
Total B	23761 ± 62	17360 ± 51	14671 ± 45	3101 ± 16
$t\bar{t}H$ (S)	84.0 ± 0.7	$140.9 \pm$	$0.9\,232 \pm 2$	53.1 ± 0.5
$S/\sqrt{B}$	0.55	1.07	1.91	0.95
Data	22900	17097	14057	3332

background events are expected while SM calculations predict an amount of 510 ± 4 events for the $t\bar{t}H$ signal process, resulting in a combined $S/\sqrt{B}$ ratio of 2.10 ± 0.02. Throughout all categories, the ratio of the number of signal events to that of the irreducible $t\bar{t}+b\bar{b}$ background is approximately 10%. This observation implies that the applied selection criteria act similar on both processes, expressing the demand for sophisticated methods that achieve further separation. Simultaneously, $t\bar{t}+$lf processes yield the largest background contribution in all categories, although with degrading influence at higher jet multiplicities, ranging from ~66% for single-lepton events with four jets, to ~52% for five jets, and ~36% for six or more jets. The composition of topologically similar $t\bar{t}+b\bar{b}$ events, rate-dominating $t\bar{t}+$lf contributions, and intermediate configurations due to $t\bar{t}+b/\bar{b}$, $t\bar{t}+2b$, and $t\bar{t}+c\bar{c}$ processes emphasizes the choice of the process-based multi-class classification strategy developed in this analysis (cf. Sect. 4.2.2).

A total of 57386 data events passes the selection. In general, uncertainties on the number of measured events are estimated using Poisson statistics, which yields an uncertainty of ± 240 events. In the instance of small statistics, uncertainties are modeled asymmetrically following the Neyman construction of Garwood owing the possibility of under-coverage effects [58]. With a total difference between observed and expected event yields of -4%, the overall agreement is reasonably good. The compatibility can be further quantified by defining the pull p as

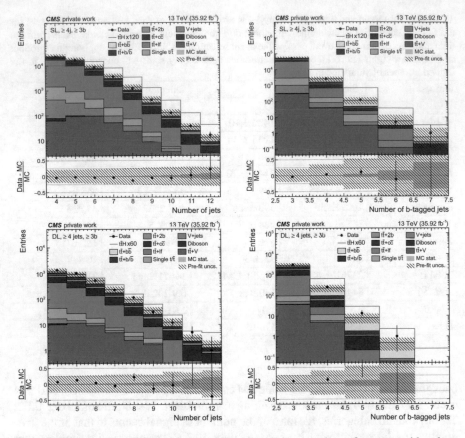

Fig. 6.6 Number of jets (left) and b-tagged jets (right) per event, shown for events with at least four jets and at least three b tags in the single-lepton (top) and in the dilepton channel (bottom). Contributions expected from background processes are stacked, while the expected $t\bar{t}H$ signal is shown superimposed and scaled by a certain factor for visibility. The filled gray areas in the ratio plots denote statistical uncertainties due to the limited number of simulated events, whereas hatched areas describe combined systematic uncertainties on expected background contributions, prior to the fit to data

$$p = \frac{N_{\text{Data}} - N_{\text{MC}}}{\sqrt{\sigma_{\text{Data}}^2 + \sigma_{\text{MC}}^2}}, \qquad (6.14)$$

which should be sufficiently small for consistent event yields given their combined uncertainties. When including the normalization uncertainty of 50% uncorrelated on all $t\bar{t}$+hf contributions (cf. Sect. 4.1), the pull in the single-lepton channel amounts to -0.29 for events with four jets, to -0.11 for five jets, to -0.20 for six or more jets, and to $+0.17$ in the dilepton channel. The combined pull is measured as -0.30, which underlines the quality of the concordance between recorded and simulated event rates.

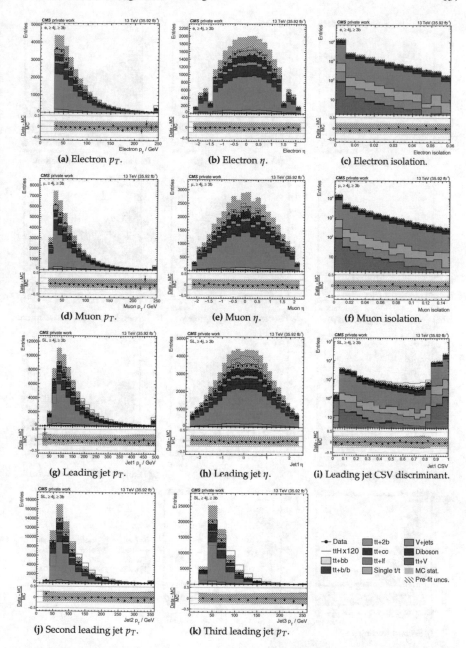

Fig. 6.7 Variable distributions in the e channel (first row), in the μ channel (second row), and in the combined single-lepton channel (bottom half). The $t\bar{t}H$ signal is shown superimposed and scaled by a factor of 120 for visibility. The filled gray areas in the ratio plots denote statistical uncertainties due to the limited number of simulated events, whereas hatched areas describe combined systematic uncertainties on expected background contributions, prior to the fit to data

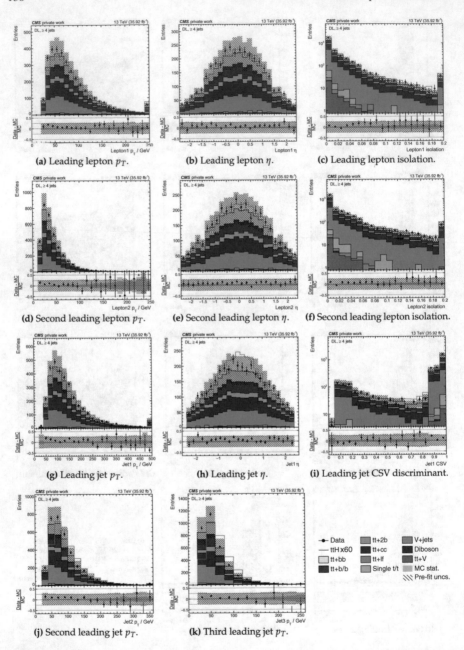

(a) Leading lepton p_T.

(b) Leading lepton η.

(c) Leading lepton isolation.

(d) Second leading lepton p_T.

(e) Second leading lepton η.

(f) Second leading lepton isolation.

(g) Leading jet p_T.

(h) Leading jet η.

(i) Leading jet CSV discriminant.

(j) Second leading jet p_T.

(k) Third leading jet p_T.

Fig. 6.8 Variable distributions in the dilepton channel. The $t\bar{t}H$ signal is shown superimposed and scaled by a factor of 60 for visibility. The filled gray areas in the ratio plots denote statistical uncertainties due to the limited number of simulated events, whereas hatched areas describe combined systematic uncertainties on expected background contributions, prior to the fit to data

Furthermore, the distributions of characteristic kinematic quantities can be compared to study the modeling of physics objects that are relevant for this analysis. A selection of distributions is presented in Figs. 6.6, 6.7, and 6.8 in the form of histograms. Contributions from expected backgrounds are stacked, while the SM prediction for the $t\bar{t}H$ signal process is superimposed and scaled by 120 and 60 in the single-lepton and dilepton channel, respectively, for the purpose of simplifying shape comparisons. Bin contents as measured in data are denoted by black markers, with vertical lines representing statistical uncertainties following the prescription of Garwood [58]. The difference between data and background simulation, divided by simulation, is visualized beneath each histogram. The filled gray areas shown in these ratio plots express statistical uncertainties due to the limited number of simulated events. Hatched areas describe combined systematic uncertainties on the expected background contributions, prior to the fit to data (cf. Sect. 8.1). They also contain a-priori normalization uncertainties on $t\bar{t}$+hf processes that are considered uncorrelated and amount to 50% (cf. Sect. 4.1).

Figure 6.6 shows the distributions of jet (left) and b tag multiplicities (right) inclusively for events with at least four jets and at least three b tags in the single-lepton (top) and dilepton channel (bottom). Within their uncertainties, all distributions exhibit a considerably good agreement in both channels. Ratios between data and simulation are flat up to high multiplicities without signs of systematic tendencies that would indicate mismodeling. For high b tag multiplicities in the dilepton channel, a slight over-fluctuation emerges, which is, however, covered by large relative uncertainties due to low event rates in the corresponding bins.

Kinematic distributions of leptons and jets in the single-lepton channel are shown in Fig. 6.7. The first and second rows present p_T, η, and isolation variables of electrons and muons, which show no sign of mismodeling. Especially the isolation variables, which compile an enhanced amount of quantities, exhibit a good compatibility between data and simulation with a flat ratio across all bins. A minor slope is visible in the transverse momentum spectra above ~ 100 GeV. However, the effect is small, well covered by the respective uncertainties, and also known from further analyses in a similar phase space [40–42]. It should be noted that the rightmost bins in histograms depicting transverse momenta also contain events that exceed the axis range to avoid visualizing extensive distribution tails. The third and fourth rows show p_T, η, and the CSV discriminant of the leading jet, and p_T distributions of the second and third leading jets. Again, ratios between data and simulation demonstrate coherent compatibility within uncertainties in all distributions. The agreement of the CSV discriminant distribution of the leading jet is notable and validates the CSV shape calibration procedure in events with multiple jets (cf. Sect. 6.4).

Figure 6.8 presents similar kinematic distributions of jets and leptons in the dilepton channel. The first two rows show p_T, η, and isolation variables of the first and second leading leptons. Both statistical uncertainties and fluctuations increase slightly with respect to the single-lepton channel, owing to the reduced amount of dilepton events. However, simulated distributions evidently succeed in modeling measured data at a reasonable level. The same holds true for lepton η and isolation variables, as well as for the η and CSV distributions of the leading jet. Spectra of transverse momenta exhibit a slope in the data-to-simulation ratios as observed by further analyses in a similar phase space [40–42]. The effect is probably related to the choice of generator tunes that affect parton showering (cf. Sect. 6.2), which is accounted for by appropriate systematic uncertainties in the construction of the statistical fit model (cf. Sect. 8.1).

Further variable distributions and a quantitative assessment of correlations between them are presented within the validation of input variables for use in deep neural networks in Sect. 7.1.

References

1. Metral E (2016) LHC: status, prospects and future challenges. In: Proceedings of the 4th annual large hadron collider physics conference, Lund, Sweden. http://cds.cern.ch/record/2265090
2. CMS Collaboration (2016) CMS public luminosity results. https://twiki.cern.ch/twiki/bin/view/CMSPublic/LumiPublicResults
3. CMS Collaboration (2017) CMS luminosity measurements for the 2016 data taking period. CMS Physics Analysis Summary CMS-PAS-LUM-17-001. http://cds.cern.ch/record/2257069
4. Petrucciani G et al (CMS Collaboration) (2015) Mini-AOD: a new analysis data format for CMS. J Phys: Conf Ser664:072052. arXiv:1702.04685 [physics.ins-det]
5. van der Meer S (1968) Calibration of the effective beam height in the ISR. CERN-ISR-PO-68-31, CERN-ISR-PO-68-31
6. Grafström P, Kozanecki W (2015) Luminosity determination at proton colliders. Prog Nucl Phys 81:97
7. Dobbs MA et al (2004) Les Houches guidebook to Monte Carlo generators for hadron collider physics. In: Proceedings of the 3rd Les Houches workshop, series C03-05-25.5, p 411. arXiv:hep-ph/0403045
8. Tanabashi M et al (2018) (Particle Data Group) The review of particle physics. Phys Rev D 98:030001
9. Frixione S, Webber BR (2002) Matching NLO QCD computations and parton shower simulations. JHEP 06:029 arXiv:hep-ph/0204244
10. Frixione S, Nason P, Oleari C (2007) Matching NLO QCD computations with parton shower simulations: the POWHEG method. JHEP 11:070 arXiv:0709.2092 [hep-ph]
11. Alwall J et al (2008) Comparative study of various algorithms for the merging of parton showers and matrix elements in hadronic collisions. Eur Phys J C 53:473 arXiv:0706.2569 [hep-ph]
12. Frederix R, Frixione S (2012) Merging meets matching in MC@NLO. JHEP 12:061 arXiv:1209.6215 [hep-ph]
13. Andersson B, Gustafson G, Ingelmann G, Sjöstrand T (1983) Parton fragmentation and string dynamics. Phys Rep 97:31
14. Sjöstrand T et al (2015) An introduction to PYTHIA 8.2. Comput Phys Commun 01:024. arXiv:1410.3012 [hep-ph]
15. CMS Collaboration (2006) CMS physics: technical design report volume 1: detector performance and software. CERN-LHCC-2006-001. http://cds.cern.ch/record/922757
16. Agostinelli S et al (The GEANT4 Collaboration) (2002) GEANT4: a simulation toolkit. Nucl Instrum Meth A 506:250. http://cds.cern.ch/record/602040
17. Allison J et al (2006) Geant4 developments and applications. IEEE Trans Nucl Sci 53:270. http://cds.cern.ch/record/1035669
18. Nason P (2004) A new method for combining NLO QCD with shower Monte Carlo algorithms. JHEP 11:040 arXiv:hep-ph/0409146

19. Alioli S, Nason P, Oleari C, Re E (2010) A general framework for implementing NLO calculations in shower Monte Carlo programs: the POWHEG BOX. JHEP 6:043 arXiv:1002.2581 [hep-ph]

20. Alwall J et al (2014) The automated computation of tree-level and next-to-leading order differential cross sections, and their matching to parton shower simulations. JHEP 07:079 arXiv:1405.0301 [hep-ph]

21. NNPDF Collaboration (2015) Parton distributions for the LHC Run II. JHEP 04:040. arXiv:1410.8849 [hep-ph]

22. Accardi A et al (2015) The PDF4LHC report on PDFs and LHC data: results from run I and preparation for run II. arXiv:1507.00556 [hep-ph]

23. Demartin F et al (2010) The impact of PDF and α_s uncertainties on Higgs Production in gluon fusion at hadron colliders. Phys Rev D 82:014002 arXiv:1004.0962 [hep-ph]

24. Collaboration CMS (2016) Event generator tunes obtained from underlying event and multi-parton scattering measurements. Eur Phys J C 76:155 arXiv:1512.00815 [hep-ex]

25. Skands P, Carrazza C, Rojo J (2013) Tuning PYTHIA 8.1: the Monash 2013 tune. Eur Phys J C 74:3024. arXiv:1404.5630 [hep-ph]

26. CMS Collaboration (2016) Investigations of the impact of the parton shower tuning in Pythia 8 in the modelling of $t\bar{t}$ at $\sqrt{s} = 8$ and 13 TeV. CMS physics analysis summary CMS-PAS-TOP-16-021. http://cds.cern.ch/record/2235192

27. Hartanto HB, Jager B, Reina L, Wackeroth D (2015) Higgs boson production in association with top quarks in the POWHEG BOX. Phys Rev D 91:094003. arXiv:1501.04498 [hep-ph]

28. Frixione S, Nason P, Ridolfi G (2007) A positive-weight next-to-leading-order Monte Carlo for heavy flavour hadroproduction. JHEP 09:126 arXiv:0707.3088 [hep-ph]

29. Alioli S, Nason P, Oleari C, Re E (2009) NLO single-top production matched with shower in POWHEG: s- and t-channel contributions. JHEP 09:111 arXiv:0907.4076 [hep-ph]

30. Artoisenet P, Frederix R, Mattelaer O, Rietkerk R (2013) Automatic spin-entangled decays of heavy resonances in Monte Carlo simulations. JHEP 03:015 arXiv:1212.3460 [hep-ph]

31. Re E (2011) Single-top Wt-channel production matched with parton showers using the POWHEG method. Eur Phys J C 71:1547 arXiv:1009.2450 [hep-ph]

32. Collaboration CMS (2019) Performance of missing transverse momentum reconstruction in proton-proton collisions at $\sqrt{s} = 13$ TeV using the CMS detector. JINST 14:P07004 arXiv:1903.06078 [hep-ex]

33. Collaboration CMS (2017) Particle-flow reconstruction and global event description with the CMS detector. JINST 12:P10003 arXiv:1706.04965 [physics.ins-det]

34. Collaboration CMS (2015) Performance of electron reconstruction and selection with the CMS detector in proton-proton collisions at $\sqrt{s} = 8$ TeV. JINST 10:P06005 arXiv:1502.02701 [physics.ins-det]

35. CMS Collaboration (2017) Electron and photon performance in CMS with the full 2016 data sample. CMS performance note CMS-DP-2017-004. http://cds.cern.ch/record/2255497

36. Collaboration CMS (2012) Performance of CMS muon reconstruction in pp collision events at $\sqrt{s} = 7$ TeV. JINST 7:P10002 arXiv:1206.4071 [physics.ins-det]

37. CMS Collaboration (2017) Muon identification and isolation efficiency on full 2016 dataset. CMS performance note CMS-DP-2017-007. http://cds.cern.ch/record/2257968

38. Bodek A et al (2012) Extracting muon momentum scale corrections for hadron collider experiments. Eur Phys J C 72:2194 arXiv:1208.3710 [hep-ex]

39. Collaboration CMS (2018) Performance of the CMS muon detector and muon reconstruction with proton-proton collisions at $\sqrt{s} = 13$ TeV. JINST 13:P06015 arXiv:1804.04528 [physics.ins-det]

40. CMS Collaboration (2017) Measurement of the $t\bar{t}$ production cross section using events in the $e\mu$ final state in pp collisions at $\sqrt{s} = 13$ TeV. Eur Phys J C 77. arXiv:1611.04040 [hep-ex]

41. Collaboration CMS (2019) Measurement of the $t\bar{t}$ production cross section, the top quark mass, and the strong coupling constant using dilepton events in pp collisions at $\sqrt{s} = 13$ TeV. Eur Phys J C 79:368 arXiv:1812.10505 [hep-ex]

42. Collaboration CMS (2019) Measurements of $t\bar{t}$ differential cross sections in proton-proton collisions at $\sqrt{s} = 13$ TeV using events containing two leptons. JHEP 02:149 arXiv:1811.06625 [hep-ex]
43. Cacciari M, Salam GP, Soyez G (2008) The anti-k_t jet clustering algorithm. JHEP 04:063 arXiv:0802.1189 [hep-ph]
44. Collaboration CMS (2017) Jet energy scale and resolution in the CMS experiment in pp collisions at 8 TeV. JINST 12:P02014 arXiv:1607.03663 [hep-ex]
45. Cacciari M, Salam GP, Soyez G (2012) FastJet user manual. Eur Phys J C 72:1896 arXiv:1111.6097 [hep-ph]
46. CMS Collaboration (2014) Performance of the particle-flow jet identification criteria using proton-proton collisions at 13 TeV for Run2016 data. CMS Analysis Note CMS-AN-17-074
47. CMS Collaboration (2018) Jet energy scale and resolution performance with 13 TeV data collected by CMS in 2016. CMS Performance Note CMS-DP-2018-028. http://cds.cern.ch/record/2622157
48. CMS Collaboration (2016) Jet energy scale and resolution performances with 13 TeV data. CMS Performance Note CMS-DP-2016-020. http://cds.cern.ch/record/2160347
49. Collaboration CMS (2013) Identification of b-quark jets with the CMS experiment. JINST 8:P04013 arXiv:1211.4462 [hep-ex]
50. Collaboration CMS (2018) Identification of heavy-flavour jets with the CMS detector in pp collisions at 13 TeV. JINST 13:P05011 arXiv:1712.07158 [physics.ins-det]
51. ATLAS Collaboration (2016) Measurement of the inelastic proton-proton cross section at $\sqrt{s} = 13$ TeV with the ATLAS detector at the LHC. Phys Rev Lett 117:182002. arXiv:1606.02625 [hep-ex]
52. Collaboration CMS (2018) Measurement of the inelastic proton-proton cross section at $\sqrt{s} = 13$ TeV. JHEP 07:161 arXiv:1802.02613 [hep-ex]
53. CMS Collaboration (2017) Muon HLT performance in 2016 data. CMS Performance Note CMS-DP-2017-056. http://cds.cern.ch/record/2297529
54. Cousins RD, Hymes KE, Tucker J (2010) Frequentist evaluation of intervals estimated for a binomial parameter and for the ratio of poisson means. Nucl Instrum Meth A 612:388 arXiv:0905.3831 [physics.data-an]
55. CMS Collaboration (2016) Performance of muon reconstruction including alignment position errors for 2016 collision data. CMS Performance Note CMS-DP-2016-067. http://cds.cern.ch/record/2229697
56. Collaboration CMS (2014) Search for the associated production of the Higgs boson with a top-quark pair. JHEP 09:087 arXiv:1408.1682 [hep-ex]
57. Cacciari M, Salam GP (2008) Pileup subtraction using jet areas. Phys Lett B 659:119 arXiv:0707.1378 [hep-ph]
58. Garwood F (1936) Fiducial limits for the poisson distribution. Biometrika 28:437

Chapter 7
Event Classification

As outlined in the measurement strategy in Sect. 4.2.2, the event categorization procedure employed in this analysis is composed of two stages. In the first stage, events are categorized according to the number of leptons and jets, which results in four categories, i.e., single-lepton (SL) events with four ($4j$), five ($5j$), or six or more jets ($\geq 6j$), and dilepton (DL) events with at least four jets ($\geq 4j$). Necessary event selection criteria are discussed in Sect. 6.3. Unlike some other analyses that involve single top or top quark pair production, the b tag multiplicity does not serve as a categorization criterion.

As the search for $t\bar{t}H$ production is background-dominated, its sensitivity strongly depends on the ability to separate and constrain background processes individually. Therefore, the second stage of event categorization is based on a multi-class classification approach using deep neural networks (DNNs). A separate DNN is trained and optimized for each of the four categories mentioned above. Each event is assigned a score, interchangeably called "probability" in the following, that describes its compatibility with either the $t\bar{t}H$ signal process or one of the five $t\bar{t}+X$ subprocesses, namely $t\bar{t}+b\bar{b}$, $t\bar{t}+b/\bar{b}$, $t\bar{t}+2b$, $t\bar{t}+c\bar{c}$, and $t\bar{t}+lf$ (cf. Sect. 2.2.4). The final category is then determined by the most probable process, i.e., the process that received the highest probability by the network. In addition, the six output values assigned to each event provide a high degree of separation between the involved processes. Hence, distributions of these DNN discriminants are utilized in the signal extraction procedure (cf. Sect. 4.2.1). The process categorization is summarized in Fig. 4.2.

This section presents the construction, validation, and evaluation of the employed deep neural networks, and is organized as follows. The set of neural network input variables is discussed in the beginning. As DNNs are potentially able to exploit complex relations between these variables, a method for validating higher-order variable correlations is introduced, which is based on saturated goodness-of-fit tests as outlined in Sect. 5.3.3. Second, the neural network architecture and the specially

© The Author(s), under exclusive license to Springer Nature Switzerland AG 2021
M. Rieger, *Search for t̄tH Production in the H → bb̄ Decay Channel*, Springer Theses,
https://doi.org/10.1007/978-3-030-65380-4_7

tailored, two-stage training procedure are presented. As the DNN discriminant distributions are used in the signal extraction procedure, an optimization algorithm is applied to ensure a minimal amount of events in each bin. This binning optimization is described in the third section. Thereafter, training results are discussed in terms of process classification accuracies and robustness against overtraining effects. The section concludes with the presentation of DNN discriminant distributions after the two-staged event categorization.

7.1 Input Variables and Validation of Correlations

This section introduces the variables that are considered as inputs for the DNNs to perform a multi-class classification. The set of variables is explained first, followed by the validation of correlations between all pairs of variables. Only variables that pass a certain selection criterion within the validation procedure are used as neural network inputs.

Input Variables

The set of potential DNN inputs comprises up to 64 variables that result from single object quantities, such as transverse momenta of leptons and jets, or combined variables that incorporate information of multiple physics objects, such as angular distributions between momentum vectors. They may vary between the single-lepton and dilepton channel, owing to different compositions of the final state, and can be subdivided into six groups.

- **Single kinematic quantities** contain transverse momenta and pseudorapidities of one (SL) or two leptons (DL) and the first four jets.[1] In the context of neural networks, they are commonly referred to as low-level variables.
- **Compound jet variables** are computed for three collections of jets, namely all selected jets, b-tagged, and non-b-tagged jets. They consist of the scalar sum of all transverse jet momenta H_T, the minimum and maximum distances $\Delta R = \sqrt{\Delta \eta^2 + \Delta \phi^2}$ of all jet pairs, the minimum distance ΔR between a jet and the leading lepton, and the mean distance $\Delta \eta$ between jet pairs. It should be noted that the latter variable is only determined for b-tagged jets.
- **Mass variables** are computed for pairs of particles using their four-vector information via $m = \sqrt{(E_1 + E_2)^2 - (p_1 + p_2)^2}$. They comprise the mean mass of all pairs of selected and b-tagged jets, the mass of the leading lepton and the closest b-tagged jet as defined by ΔR, and the mass of the b-tagged jet pair that is closest to 125 GeV. For dilepton events, the mass m_{ll} of the two leptons is considered additionally.
- **Information from b-tagging** is compiled into 15 variables. They contain CSV discriminants of the first four jets[1], as well as the second highest CSV value of all selected jets. Furthermore, the minimum, maximum, and mean CSV values are

---

[1] Sorted by transverse momentum in descending order.

considered, obtained separately for the three collections of jets mentioned above (see **Compound jet variables**).

- **Event shape variables** transform four-vectors of final-state particles into a single quantity that expresses geometrical correlations between them. The second, third, and fourth Fox-Wolfram moments, i.e., compositions of spherical harmonics [1, 2], are calculated for all jets. Further geometrical event shape quantities are determined separately for the three jet collections mentioned above (see **Compound jet variables**). Centrality describes the ratio between the sums of transverse momenta and energies of involved jets, $\sum_j p_{T,j} / \sum_j E_j$. Sphericity S, transverse sphericity, S_T, and aplanarity A are defined through the eigenvalues λ_1, λ_2, and λ_3 of the momentum tensor M as [3]

$$M = \frac{1}{\sum_j p_j^2} \cdot \sum_j \vec{p}_j \cdot \vec{p}_j^T \qquad (7.1)$$

$$\Rightarrow \quad S = \frac{3}{2}(\lambda_2 + \lambda_3), \qquad S_T = \frac{2\lambda_2}{\lambda_1 + \lambda_2}, \qquad A = \frac{3}{2}\lambda_1. \qquad (7.2)$$

- **A matrix element method (MEM) discriminant** [4–6] is constructed specifically for the purpose of separating contributions from $t\bar{t}H$ signal and $t\bar{t}+b\bar{b}$ background events [7]. It is based on the ratio

$$P_{S/B} = \frac{\omega_S(x)}{\omega_S(x) + \kappa \cdot \omega_B(x)} \qquad (7.3)$$

of two probability density functions $\omega_{S/B}$ that describe the likelihood for a reconstructed event to either originate from a $t\bar{t}H$ or a $t\bar{t}+b\bar{b}$ event. κ is set to 0.1 as a result of an optimization procedure. The densities $\omega_{S/B}$ depend on the four-vectors x of all selected final-state objects, i.e., jets and the leading lepton, that are used to compute leading-order scattering amplitudes for gluon-gluon induced $t\bar{t}H$ or $t\bar{t}+b\bar{b}$ production. Amplitudes are integrated over unknown and poorly measured particle quantities, whereby detector resolution effects are taken into account via transfer functions. This procedure is repeated for all permutations emerging from the ambiguous association between reconstructed jets and partons on generator-level. Eventually, the permutation with the highest probability is selected and the resulting probability density values ω_S and ω_B yield the MEM discriminant $P_{S/B}$ through Eq. 7.3. High discriminant values signify compatibility with the signal hypothesis. Details on the algorithm in the context of $t\bar{t}H$ analyses are presented in Ref. [7].

In order to reduce the amount of computationally demanding phase space integrations, information from b-tagging is exploited [7]. The four bottom quarks from decays of top quarks and the Higgs boson are only assigned to one of the four jets that most likely originate from b hadron decays. Therefore, two likelihoods are constructed that express the compatibility of four jets with the hypotheses of either two ($\mathcal{L}_2$) or four bottom quarks ($\mathcal{L}_4$). The likelihoods interpret the CSV discriminant values as probabilities with flavor-dependent density functions, estimated from simulation. The four jets that maximize the "b-tag likelihood ratio" (BLR), defined as $\mathcal{L}_4/(\mathcal{L}_2 + \mathcal{L}_4)$, are selected for associations to bottom quarks within permutations. The maximum BLR value yields a well separating variable itself and is thus incorporated as an input variable. However, its range of values is rather broad due to its nature of a likelihood ratio. For this reason, the transformed ratio $\log(b/(1 - b))$ is investigated in addition.

Whereas the two BLR variables are calculated for all events, the MEM discriminant is only computed in the single-lepton channel.

Owing to the jet and b tag selection strategy in the dilepton channel that requires at least four jets with at least three of them having a b tag, the amount of events with a high number of non-b-tagged jets is rather small. Therefore, the computation of some of the listed variables are omitted for the collection of non-b-tagged jets. A selection of input variable distributions is shown in Appendix A.

Variable Validation

The deep neural networks employed in this analysis are trained and optimized using simulated events, and subsequently evaluated on recorded collision data to infer predictions on underlying physics processes. The validity of neural network models is typically tested by comparing predicted distributions obtained for simulated and measured data events. In some occasions, discrepancies may arise due to overtraining, i.e., a network is incapable of generalizing on underlying probability distributions in data and rather becomes sensitive to non-representative features present in training data. Overtraining can be efficiently suppressed by reducing the network capacity, by increasing the amount of statistics available for training, or by applying further methods that optimize the loss minimization procedure (cf. Sect. 5.2.3).

Another source for potential discrepancies emerges from differences in the modeling of utilized input variables. The selection of well modeled variables is often premised upon comparisons of one-dimensional distributions between data and simulation. However, as deep neural networks potentially exploit deep correlations between particular input variables, one-dimensional distribution comparisons are not sufficient. Possible biases in the modeling of simulated events might only become apparent when scrutinizing correlations among variables.

Therefore, an extensive validation method is deployed that explicitly evaluates the modeling of variables correlations. The algorithm is defined as follows.

1. Two-dimensional histograms are created for all pairs of considered input variables, separately for simulated and recorded data events in each of the four analysis categories. The number of bins is adjusted so that each bin exhibits a reasonable number of events. Uncertainties are propagated accordingly and comprise statistical uncertainties due to the limited amount of simulated events, statistical uncertainties on the number of data events per bin using Poisson statistics following the prescription of Garwood [8], and uncertainties on production cross sections of all involved processes (Table 2.4). The latter entails additional, uncorrelated normalization uncertainties on $t\bar{t}$+hf processes that amount to 50% (cf. Sect. 4.1).

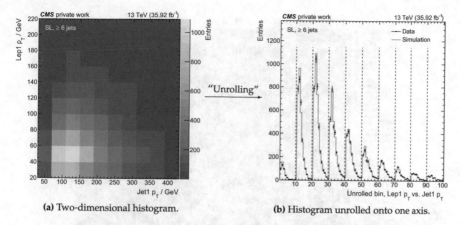

(a) Two-dimensional histogram. (b) Histogram unrolled onto one axis.

Fig. 7.1 Exemplary construction of an unrolled histogram of two input variable candidates for single-lepton events with at least six jets. Starting from a two-dimensional distribution (**a**), bin contents are transferred in column-row order to a one-dimensional histogram (**b**), which is shown including measured data events. These histograms are used to perform a fit to data and to extract p-values following a saturated goodness-of-fit test (cf. Sect. 5.3.3)

2. The two-dimensional histograms are "unrolled" to create one-dimensional distributions. A visualization of this procedure is shown in Fig. 7.1 for the transverse momenta of the leading lepton and the leading jet in single-lepton events with at least six jets.
3. A saturated goodness-of-fit test (cf. Sect. 5.3.3) is performed for all unrolled, one-dimensional histograms by fitting simulated distributions to data under consideration of the statistical model mentioned above. The fit is performed using the `combine` tool (v7.0.1), which is developed by the CMS experiment [9]. The p-value of each fit, i.e., the probability of finding a value of the test statistic that is equal to or higher than the observed value, is retained and serves as a compatibility measure for the correlation of a pair of variables. A high value signifies a good compatibility between simulation and data.
4. An iterative filtering method is used to derive selection decisions for individual variables:

   (a) Determine variables that are attributed p-values below 0.3 in combination with more than 50% of the other variables. Variables selected by this criterion are regarded as insufficiently well modeled.
   (b) Given the variables selected in step (**a**), exclude the one with the lowest sum of p-values with respect to all other variables from further considerations.
   (c) Repeat the procedure at step (**a**). If no variable is found, the filtering algorithm is terminated and remaining variables are used as neural network inputs.

Although only correlations of pairs of variables are evaluated, the method also examines higher-order correlations in data due to the composition of the set of potential

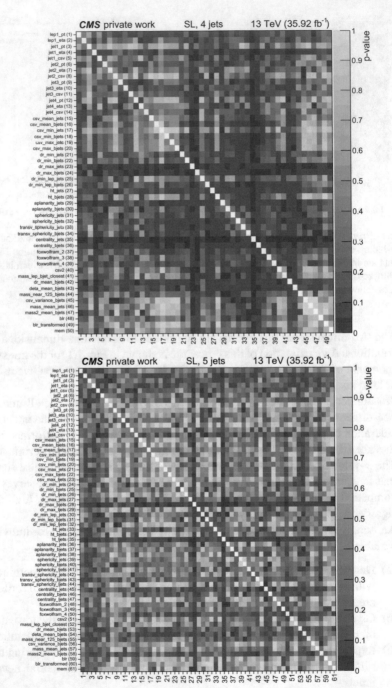

Fig. 7.2 Visualization of *p*-values derived from saturated goodness-of-fit tests. The diagrams show results for single-lepton events with four jets (top) and five jets (bottom). High *p*-values denote well modeled correlations between variables. Numerical values on the horizontal axis refer to corresponding variables on the vertical axis

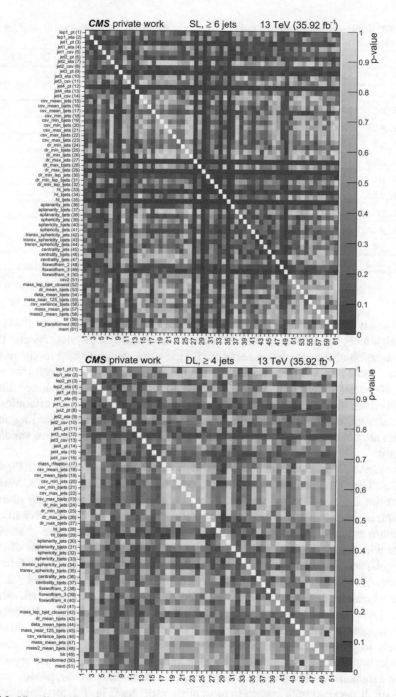

Fig. 7.3 Visualization of p-values derived from saturated goodness-of-fit tests. The diagrams show results for single-lepton events with at least six jets (top) and for dilepton events with at least four jets (bottom). High p-values denote well modeled correlations between variables. Numerical values on the horizontal axis refer to corresponding variables on the vertical axis

Table 7.1 Input variable candidates and their results of the validation procedure, divided into the three categories of single-lepton events and one dilepton category

Name	Description	SL $4j$	SL $5j$	SL $\geq 6j$	DL $\geq 4j$
jet1_pt	Jet1 p_T	✓	✓		✓
jet1_eta	Jet1 η		✓	✓	✓
jet2_pt	Jet2 p_T		✓		
jet2_eta	Jet2 η	✓	✓	✓	✓
jet3_pt	Jet3 p_T		✓		
jet3_eta	Jet3 η	✓	✓	✓	
jet4_pt	Jet4 p_T	✓	✓		✓
jet4_eta	Jet4 η	✓	✓	✓	
lep1_pt	Lepton1 p_T		✓	✓	✓
lep1_eta	Lepton1 η	✓		✓	✓
lep2_pt	Lepton2 p_T				
lep2_eta	Lepton2 η				✓

variables. It contains low-level observables from particle quantities as well as complex high-level variables that compile information of multiple physics objects. Thus, the validation of the correlation between, for example, the transverse momentum of a lepton and a Fox-Wolfram moment computed for all selected jets covers the modeling of several simulated variables at once. Furthermore, it should be noted that, as described above, the fitting procedures only incorporate systematic uncertainties that affect normalizations of expected background contributions. Shape-changing uncertainties are not considered which complies with a relatively conservative validation strategy.

The obtained p-values are shown separately for each analysis category in Figs. 7.2 and 7.3. They are mirrored along the diagonal for visualization purposes. Variable names are indicated on the vertical axis in association with increasing numbers that correspond to bins on the horizontal axis. After applying the iterative filtering algorithm as explained above, 34 variables remain in the single-lepton channel for events with four jets, 42 variables in case of five jets, 41 variables in case of six or more jets, and 39 variables in the dilepton channel for events with at least four jets. The sets of selected variables per category are presented in Tables 7.1 and 7.2.

Two variables that are rejected by the validation procedure are exemplarily shown in Fig. 7.4 for single-lepton events with four jets. The maximum distance ΔR between two b-tagged jets (left) shows a systematic trend in the data-to-simulation ratio which is likely to propagate also to correlations with other variables, eventually causing the rejection decision. In contrast, the pseudorapidity of the leading jet appears to be well modeled at first glance. The goodness-of-fit tests, however, predominantly yield small p-values with respect to most other variables (Fig. 7.2, top diagram, row 4). A validation procedure based on comparisons of one-dimensional distributions

Table 7.2 Input variable candidates and their results of the validation procedure, divided into the three categories of single-lepton events and one dilepton category

Name	Description	SL 4j	SL 5j	SL $\geq 6j$	DL $\geq 4j$
ht_jets	H_T of jets		✓		
ht_bjets	H_T of b-tagged jets	✓	✓	✓	
dr_min_jets	Min. ΔR between jets	✓	✓	✓	✓
dr_min_bjets	Min. ΔR between b-tagged jets	✓	✓	✓	✓
dr_max_jets	Max. ΔR between jets			✓	✓
dr_max_bjets	Max. ΔR between b-tagged jets			✓	✓
dr_mean_bjets	Mean ΔR between b-tagged jets		✓	✓	✓
dr_min_lep_jets	Min. ΔR between leptons and jets	✓	✓		
dr_min_lep_bjets	Min. ΔR between leptons and b-tagged jets		✓	✓	
deta_mean_bjets	Mean $\Delta \eta$ between b-tagged jets			✓	✓
mass_dilepton	Dilepton mass				✓
mass_lep_bjet_closest	Mass of lepton and closest b-tagged jet	✓	✓	✓	✓
mass_mean_jets	Mean mass of jet pairs	✓	✓	✓	✓
mass2_mean_bjets	Mean squared mass of b-tagged jet pairs	✓		✓	✓
mass_near_125_bjets	Mass of b-tagged jet pairs, closest to m_H		✓	✓	✓
jet1_csv	Jet1 CSV	✓	✓	✓	
jet2_csv	Jet2 CSV	✓	✓	✓	
jet3_csv	Jet3 CSV	✓	✓	✓	✓
jet4_csv	Jet4 CSV	✓		✓	
csv2	Second highest CSV	✓	✓	✓	✓
csv_min_jets	Min. CSV of jets	✓	✓	✓	✓
csv_min_bjets	Min. CSV of b-tagged jets	✓	✓	✓	✓
csv_max_jets	Max. CSV of jets	✓	✓	✓	✓
csv_max_bjets	Max. CSV of b-tagged jets	✓	✓	✓	✓
csv_mean_jets	Mean CSV of jets	✓	✓	✓	✓
csv_mean_bjets	Mean CSV of b-tagged jets	✓	✓	✓	✓
csv_variance_bjets	CSV variance of b-tagged jets	✓	✓	✓	✓
aplanarity_jets	Aplanarity of jets		✓	✓	✓
aplanarity_jets	Aplanarity of b-tagged jets	✓	✓	✓	✓
centrality_jets	Centrality of jets			✓	✓
centrality_jets	Centrality of b-tagged jets			✓	✓
sphericity_jets	Sphericity of jets	✓	✓	✓	
sphericity_jets	Sphericity of b-tagged jets			✓	✓
transv_sphericity_jets	Transverse sphericity of jets	✓	✓	✓	✓
transv_sphericity_bjets	Transverse sphericity of b-tagged jets	✓	✓	✓	✓
foxwolfram_2	Fox-Wolfram moment 2		✓		✓
foxwolfram_3	Fox-Wolfram moment 3	✓	✓		
foxwolfram_4	Fox-Wolfram moment 4	✓		✓	✓
blr	b-tag likelihood ratio	✓	✓	✓	✓
blr_transformed	Transformed b-tag likelihood ratio	✓	✓	✓	✓
mem	Matrix element method discriminant	✓	✓	✓	

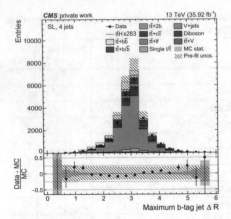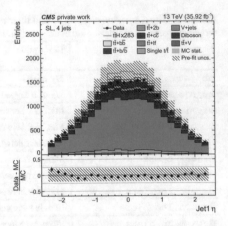

Fig. 7.4 The maximum distance ΔR between two b-tagged jets (left) and the pseudorapidity of the leading jet (right) for events in the single-lepton channel with four jets. Both variables are rejected by the validation procedure, as visible in rows 24 and 4 in the top diagram in Fig. 7.2

between data and simulation would prove ineffective under such circumstances. Distributions of selected variables for single-lepton events with at least six jets are shown in Appendix A.

7.2 Neural Network Architecture and Training

This section describes the architecture and training procedure of the developed neural networks for event categorization and process discrimination. Their task is to perform a multi-class classification with six output units whose values represent the compatibility of an event with either the $t\bar{t}H$ signal process or one of the five $t\bar{t}+X$ subprocesses, i.e., $t\bar{t}+b\bar{b}$, $t\bar{t}+b/\bar{b}$, $t\bar{t}+2b$, $t\bar{t}+c\bar{c}$, or $t\bar{t}+$lf (cf. Sect. 2.2.4). A separate network model is trained for each of the four analysis categories with input variables passing the validation procedure as presented in the previous section and summarized in Tables 7.1 and 7.2. The technical implementation is based on the TENSORFLOW graph computation library [10] (cf. Sect. 3.4.3).

Two-Staged Training Procedure

Network architectures and training procedures are defined in two stages, which are motivated as follows. When considering a $t\bar{t}H$ event, the two jets originating from the Higgs boson decay are not necessarily contained in the set of selected jets owing to requirements on identification criteria and jet kinematics (cf. Sect. 6.3.5). The probability to discard at least one jet associated to the Higgs boson decay can be estimated from simulations through a matching of partons of the hard interaction process at generator-level to reconstructed, selected jets using a ΔR distance criterion. For single-lepton events with at least six jets, the probability amounts to $\sim$27%, which conversely implies that the Higgs boson is fully reconstructable in only $\sim$73% of

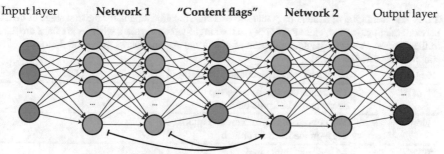

Fig. 7.5 Visualization of the two-stage neural network architecture. "Network 1" is trained to reproduce content flags in the first stage of network training. In the second training stage, it is extended by "Network 2" for the prediction of actual physics process classes. In order to avoid an artificial bottleneck, all layers of "Network 1" are fully connected to the first layer of "Network 2". It should be noted that the employed network architecture varies between the four analysis categories. The full list of network hyperparameters is presented in Table 7.4

$t\bar{t}H$ events. For events with five (four) jets, the probability even decreases to $\sim$64% ($\sim$52%). As variables related to the Higgs boson are likely to exhibit sensitive distinction features with respect to $t\bar{t}$ background contributions, the topology of signal events that miss these characteristic features might appear to neural networks as being background-like. To some extent, these topological differences between events of the same class, i.e., the same physics process, can be regarded as a source of noise during training. As the networks also distinguish between individual backgrounds, the same reasoning applies for the appearance of all involved background processes.

The developed architecture is intended to mitigate this circumstance. As depicted in Fig. 7.5, it consists of two networks which are trained sequentially. First, a fully-connected network ("Network 1") is trained to predict so-called "content flags". A flag is either zero or one and describes whether a specific particle, such as the Higgs boson, is potentially reconstructable given the set of selected jets in an event. The value of a flag for the utilization as a training target is inferred from generator information through a matching to reconstructed jets as explained above. Employed flags are summarized in Table 7.3 and vary between single-lepton and dilepton channels. The determination of additional bottom quarks expressed by the last two flags follows from a matching of generator-level jets to reconstructed jets and obeys the definition of $t\bar{t}$ subprocesses as discussed in Sect. 2.2.4.

After the training process converged, the model is extended by a second network as shown in Fig. 7.5 ("Network 2"). Its output units represent the prediction of the actual multi-class classification, reflected by values that are either one to signify one of the six involved physics processes, or zero otherwise. In order to avoid an artificial bottleneck caused by the intermediate layer expressing content flags, all layers of the first network are fully connected to the first layer of the second network through additional trainable parameters. This way, inner representations in data examined by

Table 7.3 Definition of content flags that serve as classification targets for the first stage of neural network training. In total, eight flags apply to the single-lepton channel, whereas six flags are used in the dilepton channel. The determination of additional bottom quarks obeys the definition of $t\bar{t}$ subprocesses (cf. Sect. 2.2.4)

Flag	Detectable/reconstructable particle(s)	SL	DL
"bhad"	The bottom quark from the top quark decay in the hadronic branch	✓	
"blep"	The bottom quark from the top quark decay in the leptonic branch	✓	
"lj"	At least one light quark from the hadronically decaying W boson	✓	
"Whad"	Both light quarks from the hadronically decaying W boson	✓	
"bt"	At least one bottom quark from the top quark decays		✓
"bbt"	Both bottom quarks from the top quark decays		✓
"bH"	At least one bottom quark from the Higgs boson decay ($t\bar{t}H$ only)	✓	✓
"H"	Both bottom quarks from the Higgs boson decay ($t\bar{t}H$ only)	✓	✓
"b"	At least one additional bottom quark ($t\bar{t}$ only)	✓	✓
"bb"	At least two additional bottom quarks ($t\bar{t}$ only)	✓	✓

the first network can be incorporated by the subsequent network during the second stage of the training procedure. Here, parameters of the first network that are initially optimized for the prediction of content flags are allowed to be updated through backpropagation. The first training stage can therefore be perceived as a partial pre-training. The improvement of this two-stage approach on the classification accuracy is described below in Sect. 7.4.

The available amount of simulated $t\bar{t}H$ and $t\bar{t}+X$ events is divided into three statistically independent datasets. 50% of the events are retained for the prediction of expected signal and background contributions within the signal extraction procedure of this analysis. The neural network training process is performed with 30% of the events. Both the network architecture and the training process are subject to a variety of hyperparameters whose optimal values are selected through a grid-based scanning procedure. In practical terms, this procedure constitutes another optimization process, which is validated using the remaining 20% of events.

Discrepancies observed between metrics on training and validation events permit the identification of overtraining. For this purpose, the value of the loss function to be minimized as well as the overall process classification accuracy are monitored over the course of a training. The latter is defined as the ratio of the number of events whose physics process is correctly classified to the total number of events. The degree of overtraining is quantified as the difference between the prediction accuracy obtained on the training dataset with respect to the validation dataset. Trainings that exhibit a difference larger than 1% are terminated and excluded from further considerations. Otherwise, trainings are stopped if the validation loss yields no significant improvement over the previous 100 epochs or after a maximum of 500 epochs.

Another adjustment of the training procedure stems from imbalanced expected contributions between the involved processes. Whereas $t\bar{t}+$lf events account for the majority of expected events, processes such as $t\bar{t}+b\bar{b}$, $t\bar{t}+b/\bar{b}$, and also $t\bar{t}H$ yield smaller contributions. Consequently, this imbalance causes the relevance of $t\bar{t}+$lf events to be overestimated in the training objective. As shown in Table 6.8, the contributions of all $t\bar{t}+$hf processes exceed the expected number of signal events by far, so that the additional normalization uncertainties of 50% have a strong influence on the signal extraction procedure (cf. Sect. 4.1). Therefore, event weights are introduced to equalize the importance of each of the six physics processes during training. They are determined such that the sum of weights is equal per class, while taking into account the original event weights resulting from Eqs. 6.1 and 6.7.

Architectures and Hyperparameters

Per category, the best set of hyperparameters is selected based on the maximum combined classification accuracy, with equal influence of each physics process as explained above. In all four deployed neural network architectures, the first network is comprised of two hidden layers with 100 exponential linear units each. The output layer, representing predictions of content flags, consists of eight units for networks in the single-lepton channel, and six units in the dilepton channel. A logistic sigmoid activation function is applied to constrain output values to a range between zero and one (Fig. 5.3). The objective function of the first training stage is given by a multi-label cross-entropy loss. The second network incorporates either one or two hidden layers with 100 exponential linear units each. The six units in the output layer are subject to a softmax activation function, which normalizes the sum of outputs to unity (Eq. 5.5). A cross-entropy loss constitutes the objective function of the extended network.

In both training stages, the loss is minimized using the ADAM algorithm (cf. Sect. 5.2.3) with a batch size of 5000 and an initial learning rate of $\eta = 10^{-4}$. For networks in the single-lepton (dilepton) channel, overtraining is suppressed by applying L2 regularization with a weight of 10^{-5} (10^{-3}) and a simultaneous random unit dropout with a "keep-probability" of 70% (90%). Input variables are normalized so that individual distributions are centered around zero with a standard deviation of one. A summary of selected hyperparameters is presented in Table 7.4.

Output Distributions

The trainings that yield the best combined classification accuracies are selected for performing the event categorization and process discrimination. Output distributions of the network trained in the single-lepton channel with events containing six or more jets are exemplarily shown in Fig. 7.6a–f, separately for each of the six output

Table 7.4 Summary of hyperparameters of the employed deep neural networks per category. "Network 1" refers to the network that is designed to reproduce content flags in the first training stage, whereas "Network 2" denotes the network extension for the prediction of actual process classes in the second stage. Specifications such as "100 − 100" signify the number of hidden layers and the number of units per layer. The dropout column states probabilities for retaining units in the dropout algorithm. Loss minimization is performed through the ADAM algorithm (cf. Sect. 5.2.3) with a batch size of 5000 and an initial learning rate of $\eta = 10^{-4}$. The number of trainable parameters is shown in the last column

Category		# Variables	Network 1	Network 2	Dropout	L_2 weight	# Parameters
SL	$4j$	34	100 − 100	100 − 100	0.7	10^{-5}	46,014
SL	$5j$	42	100 − 100	100	0.7	10^{-5}	36,714
SL	$\geq 6j$	41	100 − 100	100 − 100	0.7	10^{-5}	47,214
DL	$\geq 4j$	39	100 − 100	100	0.9	10^{-3}	36,012

units. Here, the categorization based on the prediction of the most probable process is not applied. Corresponding distributions for events with four jets, five jets, and in the dilepton channel with at least four jets are presented in Appendix B. The neural network output in each unit is shown for the six involved processes using a logarithmic scale and normalized to unity. The output range between zero and one is divided into 50 bins, resulting in a constant bin width of 0.02. However, the ranges shown on the horizontal axes are limited to populated regions for visualizing shape differences.

Figure 7.6a presents the output of the unit that predicts the compatibility of an event with the $t\bar{t}H$ signal process, referred to as "DNN discriminant $t\bar{t}H$" in the following. The shape of the distribution of $t\bar{t}H$ events differs significantly from shapes of the $t\bar{t}$ processes. This difference is quantifiable by means of the area under the curve of the receiver operating characteristic (ROC AUC). A value close to one signifies that two normalized distributions do not overlap, whereas a value of 0.5 denotes identical shapes with 100% overlap. Thus, a high ROC AUC score is preferable when discriminating between processes. A value of 0.75 is obtained for the comparison between $t\bar{t}H$ and the sum of the five $t\bar{t}$ processes, which indicates that the employed neural network is capable of distinguishing between signal and background topologies. According to expectations based on similarities of final states, the most closely resembling output shape results from $t\bar{t}+b\bar{b}$ contributions with a corresponding ROC AUC score of 0.66. Consistent with this reasoning, the largest shape difference is observed with respect to $t\bar{t}+$lf and $t\bar{t}+c\bar{c}$ events, reflected by ROC AUC scores of 0.81 and 0.76, respectively.

Qualitatively similar statements can be inferred for the DNN discriminants of the $t\bar{t}+b\bar{b}$ unit in Fig. 7.6b and the $t\bar{t}+2b$ unit in Fig. 7.6c. The respectively matching physics process tends to be attributed larger discriminant values with respect to other processes, leading to combined ROC AUC scores of 0.69. The most similar shape results from $t\bar{t}+b/\bar{b}$ events in both instances. Interestingly, the DNN discriminant of the $t\bar{t}+2b$ unit exhibits the largest separation power compared to $t\bar{t}H$, meaning that characteristics of $t\bar{t}+2b$ events, as perceived by the neural network, possess stronger similarities to other $t\bar{t}$ processes than to $t\bar{t}H$ events.

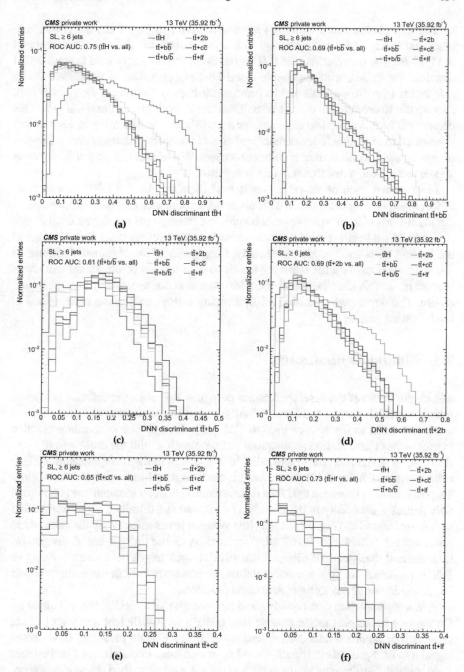

Fig. 7.6 Distributions of the six output units of the deep neural network for single-lepton events with at least six jets, namely $t\bar{t}H$ (**a**), $t\bar{t}+b\bar{b}$ (**b**), $t\bar{t}+b/\bar{b}$ (**c**), $t\bar{t}+2b$ (**d**), $t\bar{t}+c\bar{c}$ (**e**), and $t\bar{t}+$lf (**f**). The output of each unit is shown separately for $t\bar{t}H$ and $t\bar{t}$ processes. The value range is adjusted individually per unit for visualizing shape differences. For the output units $t\bar{t}H$ (**a**), $t\bar{t}+b\bar{b}$ (**b**), and $t\bar{t}+2b$ (**d**), the shape of the corresponding physics process yields clear differences with respect to other processes, which is also reflected by the ROC AUC scores

The same holds true for the $t\bar{t}+b/\bar{b}$ discriminant in Fig. 7.6c, the $t\bar{t}+c\bar{c}$ discriminant in Fig. 7.6f, and, although to a lesser extent, for the $t\bar{t}+lf$ discriminant in Fig. 7.6f. The distributions for $t\bar{t}H$ signal events consistently tend towards small discriminant values, with reasonable discrimination regarding $t\bar{t}$ processes. While $t\bar{t}H$ and $t\bar{t}+b\bar{b}$ distributions yield a prominent shape difference in the $t\bar{t}+b/\bar{b}$ unit, they appear to converge in the $t\bar{t}+c\bar{c}$ unit, and more evidently in the $t\bar{t}+lf$ unit. This behavior is both desired and expected for a multi-class classification as each output is intended to exploit individual characteristics of the corresponding process in order to equally distinguish it from all other processes. Therefore, $t\bar{t}H$ and $t\bar{t}+b\bar{b}$ events appear alike given a discriminant that is sensitive to $t\bar{t}+lf$.

In summary, each of six processes is best identified by the DNN discriminant of its corresponding output unit. This observation demonstrates the effectiveness of the multi-class classification approach achieved through the employed deep neural networks. As a result, events are categorized by the process class that is assigned the highest output value. Simultaneously, histograms of the obtained discriminant distributions constitute the basis for the fit to data as part of the signal extraction procedure. To this end, the use of equal bin widths is not necessarily an appropriate choice. The developed algorithm for optimizing histogram binning is explained in the following section.

7.3 Binning Optimization

The output layers of the developed neural networks are subject to softmax activation functions, which transform the sum of unit inputs using an exponential function, and normalize them to the sum of outputs (Eq. 5.5). This choice is common practice in multi-class classification applications in combination with the cross-entropy loss (Eq. 5.8) as it yields sensitive and reliable derivatives to the backpropagation algorithm (term $\nabla_y L$ in Eq. 5.12). However, this configuration has two implications. First, the possible output range of the DNN discriminants between zero and one is occupied only partially, especially in the $t\bar{t}+b/\bar{b}$, $t\bar{t}+c\bar{c}$, and $t\bar{t}+lf$ units where discriminant values are below 0.5 (Fig. 7.6). This circumstance is not affected by the raise of the lower boundary to $1/6 \approx 0.167$ after application of the process-based categorization. Second, the shape of discriminant distributions might have sharply rising or falling segments due to the exponential transformations in both the numerator and the denominator of the softmax activation function.

A binning approach with equally sized bins entails two undesirable consequences. On the one hand, bins might emerge that contain an insufficient amount of either simulated or recorded data events, leading to an inadequate modeling of expected signal and background contributions and high statistical uncertainties. On the other hand, output distributions might exhibit steep yet well described shape characteristics, which are mitigated in case of inappropriately large bin widths. Thus, a binning optimization algorithm is applied and defined as follows.

Initial histograms are created with 20 equally sized bins in the interval between zero and one. Outer bins are identified that contain neither simulated nor recorded data events. Subsequently, the binning is repeated, however, limited to the interval

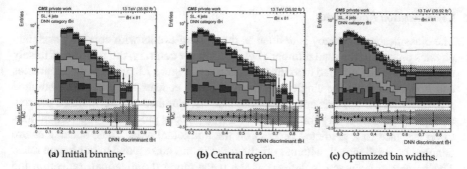

(a) Initial binning.   (b) Central region.   (c) Optimized bin widths.

Fig. 7.7 Example of the binning optimization procedure of the DNN discriminant $t\bar{t}H$. The procedure commences with 20 initial bins in the range [0, 1] (**a**). Then, outer unpopulated bins are discarded and the histogram is recreated in the resulting central region with 20 bins (**b**). Finally, bin widths are adjusted iteratively until all bins contain at least five data events (**c**)

defined by the centrally populated bins. Based on this, individual bin widths are adjusted in an iterative approach:

1. Determine bins that contain less than five events ("L" bins), or more than 20 events ("H" bins).
2. Terminate the procedure if no L bins exist.
3. Restart the procedure with a reduced number of total bins if no H bins are found.
4. Increase the width of the L bin with the lowest event yield by 0.002. Reduce the width of the widest H bin, or the bin with the largest content in case of ambiguities, by the same amount.

The procedure is repeated until no under-populated bins remain (step **2**). Figure 7.7 shows an example of the binning optimization of the DNN discriminant $t\bar{t}H$. The resulting histogram exhibits narrower bins with high statistics, while being more coarse in the tail of the distribution.

7.4 Evaluation of Training Results

The following paragraphs evaluate the results of the neural network trainings as discussed in the previous sections. First, process prediction accuracies are presented and compared between training and validation datasets to demonstrate the effectiveness of applied overtraining suppression measures. The dependence of selected network architectures and the resulting prediction accuracies on the amount of training events is evaluated thereafter. Furthermore, to estimate the potential bias towards the selected POWHEG [11] event generator, a comparison to DNN output discriminants obtained with SHERPA+OPENLOOPS [12, 13] is presented. The section closes with results of the process-based categorization scheme in terms of expected event yields and DNN discriminant distributions per category.

Classification Accuracies

The classification accuracy is defined as the ratio of events with correctly assigned physics process to the total number of events. On the contrary, the fraction of falsely classified events is referred to as the confusion rate. Figure 7.8 presents the accuracies and confusion rates for single-lepton events with at least six jets, and evaluated separately for the training and validation datasets. Numbers are normalized per row and quote the probability for an event of a true physics process on the vertical axis to be assigned to a predicted process on the horizontal axis. Thus, accuracies are denoted by the diagonal, whereas off-diagonal values correspond to confusion rates. The combined accuracy is defined as the mean along the diagonal, reflecting the equal importance among involved process classes. Slightly lower values emerge for smaller jet multiplicities and for events in the dilepton channel, which, however, lead to similar conclusions.

The comparison between the training and validation datasets suggests that measures taken for the suppression of overtraining prove effective, as differences are mostly below $\sim 1\%$. The classification accuracy for $t\bar{t}H$ events amounts to 48%, which is only surpassed by that for $t\bar{t}+\mathrm{lf}$ events with 63%. This finding expresses the significance of weights for the equalization of process classes during training. The network possesses a high accuracy for the common $t\bar{t}+\mathrm{lf}$ process, while simultaneously being able to correctly classify almost half of the rare $t\bar{t}H$ events. $t\bar{t}+b\bar{b}$ and $t\bar{t}+2b$ events are classified with accuracies of 30% and 37%, respectively, whereas $t\bar{t}+b/\bar{b}$ and $t\bar{t}+c\bar{c}$ events are mostly classified as $t\bar{t}+\mathrm{lf}$. This observation is consistent with the DNN discriminants and ROC AUC scores shown in Fig. 7.6 and indicates

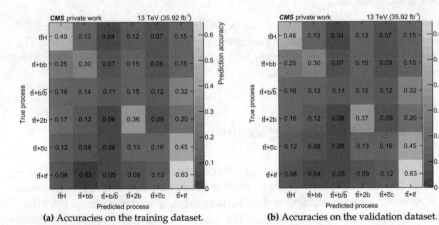

(a) Accuracies on the training dataset.   (b) Accuracies on the validation dataset.

Fig. 7.8 Row-normalized process classification accuracies and confusion rates, evaluated on the training dataset (**a**) and on the validation dataset (**b**) for single-lepton events with at least six jets. Variations between values of the two datasets are acceptably small to conclude that overtraining is sufficiently suppressed. $t\bar{t}H$, $t\bar{t}+b\bar{b}$, $t\bar{t}+2b$, and $t\bar{t}+\mathrm{lf}$ processes exhibit the highest values in the correct class. $t\bar{t}+b/\bar{b}$ and $t\bar{t}+c\bar{c}$ processes are predominantly classified as $t\bar{t}+\mathrm{lf}$, consistent with the DNN output distributions presented in Fig. 7.6

Fig. 7.9 Average process classification accuracies as a function of training statistics as evaluated on the training dataset (red) and on the validation dataset (blue) for single-lepton events with at least six jets. The dashed line marks the fraction of training statistics (~70%) which would cause an accuracy difference between training and validation events of 1%. Therefore, selected network architectures trained with the full statistics (100%) are considered stable

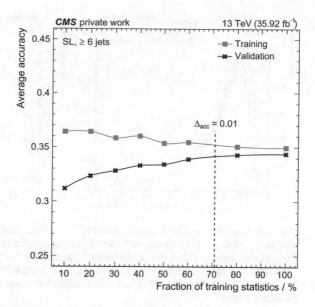

that the two latter processes are difficult to distinguish from the $t\bar{t}+$lf process given their largely similar topologies.

The enhancement provided by the two-staged neural network architecture can be assessed by repeating the training procedure and hyperparameter optimization for an architecture without content flags, consisting of up to five hidden layers with an equal amount of units per layer. For this purpose, the $t\bar{t}H$ classification accuracy is a suitable figure of merit. The best performing model results in an accuracy of 40%, opposed to the 48% obtained with the two-staged architecture. The relative improvement of 20% is considered to have a significant impact on the overall analysis sensitivity.

Dependence on Training Statistics

As explained in Sect. 7.2, trainings are terminated within the extensive hyperparameter optimization in case the average classification accuracy yields a difference larger than 1% between training and validation datasets. In addition, a supplemental study is performed to exclude that the selected network architectures entail a dependence on the amount of utilized training events. Therefore, the trainings are repeated for identical hyperparameters, however, with artificially reduced trainings statistics.

An undesired, strong dependence would manifest as a prompt increase of the difference between training and validation accuracies. However, as depicted in Fig. 7.9 for the training in the single-lepton channel with six or more jets, above a fraction of 70% of training events the difference remains well below 1%. Above that fraction, the accuracies seemingly converge to their eventual result with 100% statistics. In this light, it can be concluded that the trained neural networks provide sufficiently stable

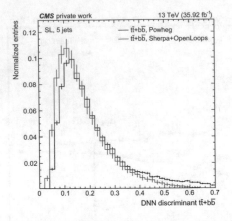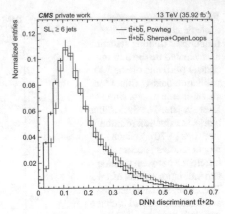

Fig. 7.10 Comparison of DNN discriminant shapes between single-lepton $t\bar{t}+b\bar{b}$ events simulated with either POWHEG [11] (black) or SHERPA [12] in combination with OPENLOOPS [13] (red). Networks are trained with events produced by POWHEG. The histograms show the DNN output of the $t\bar{t}+b\bar{b}$ unit for events with five jets (left), and the output of the $t\bar{t}+2b$ unit for events with at least six jets (right). Vertical error bars denote statistical uncertainties due to the limited number of simulated events

results. Network capacities are well balanced to methods for overtraining prevention, which achieve almost full suppression.

Dependence on Event Generator

The simulation of events that are considered for neural network trainings relies on the POWHEG (v2) [11] event generator (cf. Sect. 6.2). Naturally, the choice of a specific generator can induce a bias regarding event kinematics and the definition of $t\bar{t}$ subprocesses (cf. Sect. 2.2.4), which would propagate to the distributions of DNN discriminants. Therefore, a second event generator setup is employed and the resulting discriminant shapes are compared to estimate the significance of a potential bias.

For this study, a sample of single-lepton $t\bar{t}+b\bar{b}$ events is generated using SHERPA (v2.2.2) [12] in the four-flavor scheme in conjunction with scattering amplitudes computed by OPENLOOPS (v1.3.1) [13]. Parton showering, detector simulation and reconstruction, as well as selection requirements are identical. Selected events are evaluated by the DNNs that are trained with events generated by POWHEG. A comparison between the resulting shapes is presented in Fig. 7.10. It shows the DNN discriminants $t\bar{t}+b\bar{b}$ and $t\bar{t}+2b$ for single-lepton events with five jets and at least six jets, respectively. In central bins, the distributions agree well within statistical uncertainties, as denoted by vertical bars. For discriminant values close to zero and in the tails of the distributions, slight differences are visible which, however, do not exhibit any pronounced structure. They are well covered by background normalization uncertainties of 50% on $t\bar{t}+b\bar{b}$, $t\bar{t}+b/\bar{b}$, $t\bar{t}+2b$, and $t\bar{t}+c\bar{c}$ contributions (cf. Sect. 4.1), and by further systematic uncertainties (cf. Sect. 8.1).

Event Yields and Discriminants After Process Categorization

The expected signal and background contributions, and corresponding numbers of measured data events after the categorization by the most probable process are shown in Tables 7.5, 7.6, 7.7 and 7.8. The total numbers of simulated events agree well with the amount of data in each category. The largest discrepancies per jet multiplicity category amount to -10% for $t\bar{t}+$lf events in the SL $4j$ category, to $+7\%$ for $t\bar{t}+b/\bar{b}$ events in the SL $5j$ category, to -8% for $t\bar{t}H$ events in the SL $\geq 6j$ category, and to $+11\%$ for $t\bar{t}+b/\bar{b}$ events in the DL $\geq 4j$ category. The relative signs of discrepancies in particular process categories with respect to data are consistent across the four jet-based categories. Considering the uncertainties from limited statistics, process cross sections, and the additional normalization uncertainty of 50% on contributions from $t\bar{t}+b\bar{b}, t\bar{t}+b/\bar{b}, t\bar{t}+2b$, and $t\bar{t}+c\bar{c}$ processes (cf. Sect. 4.1), the concordance between recorded and simulated event rates is reasonably good.

Furthermore, the effect of the process-based categorization is recognizable by the relative amount of $t\bar{t}H$ events being assigned to the $t\bar{t}H$ process category, and the consequential effect on the signal-to-background ratio $S/\sqrt{B}$. The highest ratios are measured consistently in the $t\bar{t}H$ category with values ranging from 0.58 in the single-lepton channel for events with four jets, to 2.00 for events with at least six jets.

The DNN discriminant distributions after applying the process categorization are presented in Figs. 7.11, 7.12, 7.13 and 7.14. The SM prediction for the $t\bar{t}H$ signal

Table 7.5 Observed and expected event yields prior to the fit to data in the six process categories for single-lepton events with four jets (SL $4j$). Expected yields result from the total numbers of simulated event passing the event selection (Table 6.7) and applying the product of all event weights (Eq. 6.7). Stated uncertainties denote only statistical uncertainties due to the limited number of simulated events

Process	Single-lepton $4j$					
	$t\bar{t}ll$	$t\bar{t}+b\bar{b}$	$t\bar{t}+b/\bar{b}$	$t\bar{t}+2b$	$t\bar{t}+c\bar{c}$	$t\bar{t}+$lf
$t\bar{t}+b\bar{b}$	139 ± 3	191 ± 3	105 ± 3	149 ± 3	119 ± 3	133 ± 3
$t\bar{t}+b/\bar{b}$	253 ± 4	215 ± 4	326 ± 5	370 ± 5	308 ± 5	469 ± 6
$t\bar{t}+2b$	124 ± 3	77 ± 2	90 ± 2	317 ± 4	100 ± 3	134 ± 3
$t\bar{t}+c\bar{c}$	298 ± 6	232 ± 5	251 ± 5	428 ± 7	686 ± 9	1022 ± 11
$t\bar{t}+$lf	1249 ± 13	727 ± 10	1035 ± 12	1401 ± 13	2909 ± 20	8463 ± 33
Single $t/\bar{t}$	96 ± 10	116 ± 11	93 ± 10	167 ± 13	231 ± 16	304 ± 19
$V+$jets	37 ± 3	76 ± 4	27 ± 3	48 ± 3	97 ± 6	69 ± 5
Diboson	4 ± 1	5 ± 2	0.6 ± 0.6	0.9 ± 0.5	2 ± 1	4 ± 1
$t\bar{t}+V$	13.4 ± 0.9	5.6 ± 0.6	5.5 ± 0.6	11.6 ± 0.9	10.3 ± 0.8	15.8 ± 0.9
Total B	2213 ± 19	1644 ± 17	1935 ± 18	2892 ± 21	4462 ± 28	10614 ± 41
$t\bar{t}H$ (S)	27.2 ± 0.4	9.2 ± 0.2	6.8 ± 0.2	15.7 ± 0.3	8.9 ± 0.2	16.3 ± 0.3
$S/\sqrt{B}$	0.58	0.23	0.15	0.29	0.13	0.16
Data	2125	1793	2027	2896	4366	9693

Table 7.6 Observed and expected event yields prior to the fit to data in the six process categories for single-lepton events with five jets (SL 5 j). Expected yields result from the total numbers of simulated event passing the event selection (Table 6.7) and applying the product of all event weights (Eq. 6.7). Stated uncertainties denote only statistical uncertainties due to the limited number of simulated events

Process	Single-lepton 5 j					
	$t\bar{t}H$	$t\bar{t}+b\bar{b}$	$t\bar{t}+b/\bar{b}$	$t\bar{t}+2b$	$t\bar{t}+c\bar{c}$	$t\bar{t}+$lf
$t\bar{t}+b\bar{b}$	266 ± 4	410 ± 5	150 ± 3	224 ± 4	144 ± 3	228 ± 4
$t\bar{t}+b/\bar{b}$	257 ± 4	290 ± 4	321 ± 4	355 ± 5	219 ± 4	494 ± 6
$t\bar{t}+2b$	136 ± 3	128 ± 3	89 ± 2	324 ± 4	85 ± 2	184 ± 3
$t\bar{t}+c\bar{c}$	336 ± 7	341 ± 7	264 ± 6	445 ± 7	552 ± 8	1207 ± 12
$t\bar{t}+$lf	785 ± 11	647 ± 10	683 ± 9	830 ± 10	1148 ± 13	4903 ± 25
Single $t/\bar{t}$	62 ± 9	82 ± 10	45 ± 7	98 ± 10	114 ± 12	189 ± 15
$V+$jets	25 ± 2	54 ± 4	11 ± 2	34 ± 2	46 ± 3	54 ± 4
Diboson	1.4 ± 0.8	3 ± 1	–	0.4 ± 0.4	0.6 ± 0.6	3 ± 1
$t\bar{t}+V$	20 ± 1	14 ± 1	6.5 ± 0.7	17 ± 1	10.6 ± 0.9	25 ± 1
Total B	1889 ± 17	1969 ± 18	1570 ± 14	2326 ± 18	2320 ± 20	7286 ± 33
$t\bar{t}H$ (S)	52.7 ± 0.5	21.4 ± 0.3	8.0 ± 0.2	19.8 ± 0.3	10.5 ± 0.2	28.5 ± 0.4
$S/\sqrt{B}$	1.21	0.48	0.20	0.41	0.22	0.33
Data	1848	2040	1690	2299	2302	6918

Table 7.7 Observed and expected event yields prior to the fit to data in the six process categories for single-lepton events with at least six jets (SL $\geq 6j$). Expected yields result from the total numbers of simulated event passing the event selection (Table 6.7) and applying the product of all event weights (Eq. 6.7). Stated uncertainties denote only statistical uncertainties due to the limited number of simulated events

Process	Single-lepton $\geq 6j$					
	$t\bar{t}H$	$t\bar{t}+b\bar{b}$	$t\bar{t}+b/\bar{b}$	$t\bar{t}+2b$	$t\bar{t}+c\bar{c}$	$t\bar{t}+$lf
$t\bar{t}+b\bar{b}$	834 ± 8	1156 ± 9	145 ± 3	299 ± 4	17 ± 1	2.6 ± 0.4
$t\bar{t}+b/\bar{b}$	549 ± 6	575 ± 6	253 ± 4	314 ± 4	28 ± 1	3.5 ± 0.5
$t\bar{t}+2b$	306 ± 5	282 ± 4	78 ± 2	372 ± 5	9.9 ± 0.8	1.1 ± 0.3
$t\bar{t}+c\bar{c}$	1150 ± 12	998 ± 12	444 ± 7	636 ± 9	115 ± 4	16 ± 1
$t\bar{t}+$lf	1982 ± 17	1280 ± 14	916 ± 11	852 ± 11	243 ± 6	50 ± 2
Single $t/\bar{t}$	110 ± 11	146 ± 13	53 ± 8	92 ± 10	4 ± 2	3 ± 2
$V+$jets	38 ± 2	78 ± 3	10 ± 2	34 ± 2	7 ± 1	0.6 ± 0.3
Diboson	0.9 ± 0.6	0.5 ± 0.5	0.4 ± 0.4	0.4 ± 0.4	–	–
$t\bar{t}+V$	80 ± 3	58 ± 2	11.4 ± 0.9	31 ± 2	4.4 ± 0.6	0.3 ± 0.2
Total B	5050 ± 26	4575 ± 26	1911 ± 16	2629 ± 19	429 ± 8	77 ± 3
$t\bar{t}H$ (S)	142.2 ± 0.9	53.3 ± 0.5	9.5 ± 0.2	24.4 ± 0.3	2.1 ± 0.1	0.3 ± 0.0
$S/\sqrt{B}$	2.00	0.79	0.22	0.48	0.10	0.03
Data	4823	4400	1852	2484	422	76

Table 7.8 Observed and expected event yields prior to the fit to data in the six process categories for dilepton events with at least four jets (DL $\geq 4j$). Expected yields result from the total numbers of simulated event passing the event selection (Table 6.7) and applying the product of all event weights (Eq. 6.7). Stated uncertainties denote only statistical uncertainties due to the limited number of simulated events

Process	Dilepton $\geq 4j$					
	$t\bar{t}H$	$t\bar{t}+b\bar{b}$	$t\bar{t}+b/\bar{b}$	$t\bar{t}+2b$	$t\bar{t}+c\bar{c}$	$t\bar{t}+\mathrm{lf}$
$t\bar{t}+b\bar{b}$	160 ± 3	154 ± 3	88 ± 2	57 ± 2	45 ± 2	68 ± 2
$t\bar{t}+b/\bar{b}$	81 ± 2	59 ± 2	158 ± 3	80 ± 2	61 ± 2	135 ± 3
$t\bar{t}+2b$	36 ± 1	23 ± 1	50 ± 2	85 ± 2	23 ± 1	42 ± 2
$t\bar{t}+c\bar{c}$	93 ± 3	65 ± 2	108 ± 3	93 ± 2	144 ± 3	232 ± 4
$t\bar{t}+\mathrm{lf}$	63 ± 2	44 ± 2	119 ± 3	93 ± 3	133 ± 3	402 ± 5
Single $t/\bar{t}$	10 ± 3	8 ± 3	12 ± 4	8 ± 3	5 ± 3	7 ± 3
V+jets	1.6 ± 0.3	2.5 ± 0.6	1.8 ± 0.6	1.8 ± 0.3	1.9 ± 0.6	5 ± 1
Diboson	0.6 ± 0.6	–	–	–	–	0.5 ± 0.5
$t\bar{t}+V$	11 ± 1	5.5 ± 0.7	6.0 ± 0.6	4.8 ± 0.6	6.2 ± 0.7	6.6 ± 0.6
Total B	457 ± 6	361 ± 6	544 ± 7	421 ± 6	420 ± 6	898 ± 8
$t\bar{t}H$ (S)	24.2 ± 0.4	7.1 ± 0.2	6.7 ± 0.2	5.5 ± 0.2	4.9 ± 0.1	4.7 ± 0.1
$S/\sqrt{B}$	1.13	0.38	0.29	0.27	0.24	0.16
Data	509	386	622	424	436	955

process is shown superimposed and scaled to match the total amount of expected background, whose contributions are stacked. Black markers denote the number of measured events, with statistical uncertainties represented by vertical lines, following the prescription of Garwood [8]. Owing to the binning optimization procedure (cf. Sect. 7.3), the minimum amount of five events is exceeded in every bin. The ratio between data and background simulation, shifted to zero by subtracting simulations, is visualized beneath each histogram. Therein, filled gray areas express statistical uncertainties due to the limited number of simulated events, whereas hatched areas reflect the combined systematic uncertainties on the expected background contributions, prior to the fit to data, and include additional uncertainties of 50% on the production of $t\bar{t}$+hf events (cf. Sect. 4.1).

Overall, the compatibility between simulation and data is found to be reasonably good within measurement uncertainties and confirms the effect of the comprehensive validation of input variables and the conservative suppression of overtraining. Minor differences arise in regions with reduced statistics. However, they are comparably small and covered by statistical uncertainties.

The comparison between shapes of $t\bar{t}H$ and background processes in the DNN discriminants of the $t\bar{t}H$ output unit yields evident differences, where the best separation is achieved in the rightmost bins. After the fit to data, these bins could potentially contain a significant amount of $t\bar{t}H$ signal events. In the single-lepton channel, the discriminant distributions in the $t\bar{t}+b/\bar{b}$ output unit for events with four jets

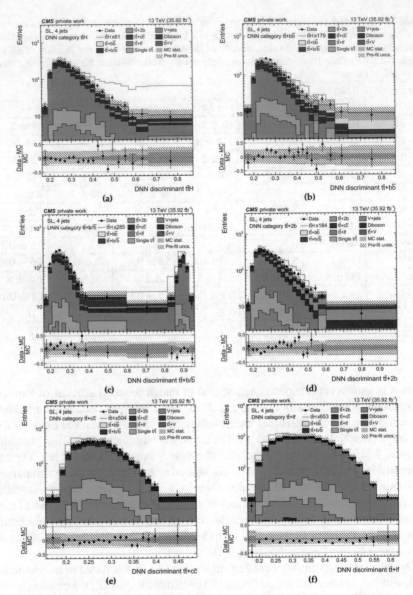

Fig. 7.11 Distributions of the six output units of the deep neural network for single-lepton events with four jets, after the additional categorization into most probable processes. For example, Fig. (**a**) shows the output distribution of the $t\bar{t}H$ unit, for events that have their highest output value in the $t\bar{t}H$ unit. The filled gray areas in the ratio plots denote statistical uncertainties due to the limited number of simulated events, whereas hatched areas describe combined systematic uncertainties on expected background contributions, prior to the fit to data

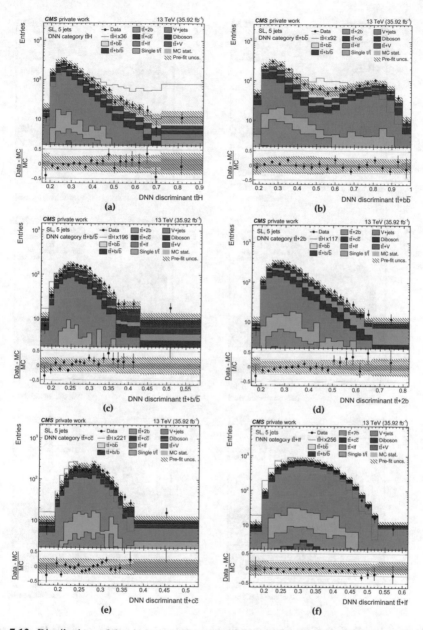

Fig. 7.12 Distributions of the six output units of the deep neural network for single-lepton events with five jets, after the additional categorization into most probable processes. For example, Fig. (**a**) shows the output distribution of the $t\bar{t}H$ unit, for events that have their highest output value in the $t\bar{t}H$ unit. The filled gray areas in the ratio plots denote statistical uncertainties due to the limited number of simulated events, whereas hatched areas describe combined systematic uncertainties on expected background contributions, prior to the fit to data

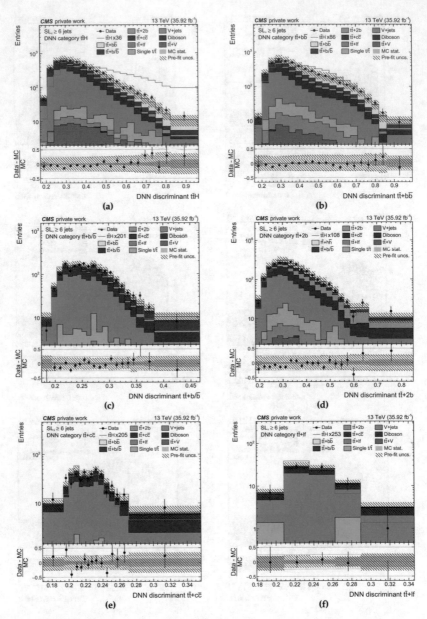

Fig. 7.13 Distributions of the six output units of the deep neural network for single-lepton events with at least six jets, after the additional categorization into most probable processes. For example, Fig. (a) shows the output distribution of the $t\bar{t}H$ unit, for events that have their highest output value in the $t\bar{t}H$ unit. The filled gray areas in the ratio plots denote statistical uncertainties due to the limited number of simulated events, whereas hatched areas describe combined systematic uncertainties on expected background contributions, prior to the fit to data

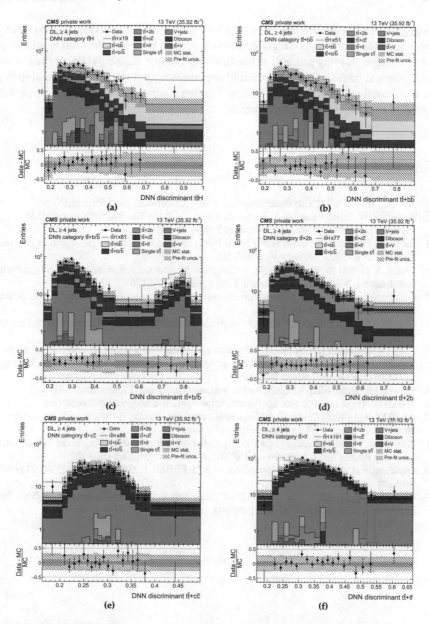

Fig. 7.14 Distributions of the six output units of the deep neural network for dilepton events with at least four jets, after the additional categorization into most probable processes. For example, Fig. (**a**) shows the output distribution of the $t\bar{t}H$ unit, for events that have their highest output value in the $t\bar{t}H$ unit. The filled gray areas in the ratio plots denote statistical uncertainties due to the limited number of simulated events, whereas hatched areas describe combined systematic uncertainties on expected background contributions, prior to the fit to data

(Fig. 7.11c) and in the $t\bar{t}+b\bar{b}$ output unit for events with five jets (Fig. 7.12b) exhibit two resolvable peaks. This structure is well described by simulation and results from the two-staged training process. When omitting the first training stage to predict the value of event content flags, comparable structures are not observed, suggesting that the compound network is able to distinguish two components in the underlying data with differing kinematic characteristics.

References

1. Fox GC, Wolfram S (1979) Event shapes in e^+e^- annihilation. Nucl Phys B 157:543
2. Bernaciak B, Buschmann MSA, Butter A, Plehn T (2013) Fox-Wolfram moments in Higgs physics. Phys Rev D 87:073014. arXiv:1212.4436 [hep-ph]
3. Barger V, Ohnemus J, Philips RJN (1993) Event shape criteria for single-lepton top signals. Phys Rev D 48:3953. arXiv:hep-ph/9308216
4. Kondo K (1988) Dynamical likelihood method for reconstruction of events with missing momentum: I. Method and toy models. J Phys Soc Jpn 57:4126
5. Collaboration DØ (2004) A precision measurement of the mass of the top quark. Nature 429:638. arXiv:hep-ex/0406031
6. Bhat PC (2011) Multivariate analysis methods in particle physics. Ann Rev Nucl Part Sci 61:281
7. CMS Collaboration (2015) Search for a standard model Higgs boson produced in association with a top-quark pair and decaying to bottom quarks using a matrix element method. Eur Phys J C 75:251. arXiv:1502.02485 [hep-ex]
8. Garwood F (1936) Fiducial limits for the poisson distribution. Biometrika 28:437
9. The CMS Higgs combine tool developers. https://cms-hcomb.gitbooks.io/combine
10. Abadi M et al (2016) TensorFlow: large-scale machine learning on heterogeneous distributed systems. Preliminary White Paper. arXiv:1603.04467 [cs.DC]
11. Frixione S, Nason P, Ridolfi G (2007) A positive-weight next-to-leading-order Monte Carlo for heavy flavour hadroproduction. JHEP 09:126. arXiv:0707.3088 [hep-ph]
12. Gleisberg T et al (2009) Event generation with SHERPA 1.1. JHEP 02:007. arXiv:0811.5622 [hep-ph]
13. Cascioli F, Maierhöfer P, Pozzorini S (2012) Scattering amplitudes with open loops. Phys. Rev. Lett. 108:111601. arXiv:1111.5206 [hep-ph]

Chapter 8
Measurement

This section presents the results of the search for $t\bar{t}H$ production conducted in this analysis. First, the statistical model is explained, which relates expected, simulated event contributions to corresponding sources of systematic uncertainties. This model constitutes the basis for extracting the signal strength modifier $\mu = \sigma/\sigma_{SM}$ in a fit to data. Results of this procedure are presented thereafter. Finally, the influence of individual sources of systematic uncertainties is evaluated.

8.1 Systematic Uncertainties

The systematic uncertainties that are relevant for the presented analysis are discussed in the following sections. The signal extraction procedure relies on the likelihood $L(\mu, \boldsymbol{\theta} \mid \boldsymbol{x})$ of μ and the nuisance parameters $\boldsymbol{\theta}$ given measured data $\boldsymbol{x}$. For the scenario of binned variable distributions, i.e., simultaneous counting experiments, the likelihood can be constructed as (cf. Sect. 5.3.1)

$$L(\mu, \boldsymbol{\theta} \mid \boldsymbol{x}) = \prod_i^{\text{bins}} \text{Poisson}\,(x_i \mid \mu \cdot s_i(\boldsymbol{\theta}) + b_i(\boldsymbol{\theta})) \tag{8.1}$$

$$\text{with} \quad b_i(\boldsymbol{\theta}) = \sum_p^{\text{processes}} \underbrace{b_{p,i}}_{\text{nominal yield}} \cdot \underbrace{\prod_n^{\dim(\boldsymbol{\theta})} \pi_{n,p,i}(\theta_n)}_{\text{uncertainty model}}, \tag{8.2}$$

where s_i, b_i, and x_i denote the expected signal, background, and measured data events in bin i, respectively. While the signal strength modifier μ scales the expected signal prediction, both s_i and b_i depend on nuisance parameters $\boldsymbol{\theta}$, which mediate the effect

© The Author(s), under exclusive license to Springer Nature Switzerland AG 2021
M. Rieger, *Search for t̄tH Production in the H → bb̄ Decay Channel*, Springer Theses,
https://doi.org/10.1007/978-3-030-65380-4_8

of systematic uncertainties as part of the uncertainty model (Eq. 8.2). The probability for a value θ_n of a specific nuisance n is described by a prior distribution $\pi_{n,p,i}$, which can depend on the particular bin i or a physics process p. In case of $\pi_{n,p,i} \to \pi_{n,p}$, the nuisance only affects the overall rate of a process. Otherwise, it also causes a shape-changing effect in the context of binned distributions. It should be noted that rate-changing nuisance can alter the composition of processes in a specific category and therefore lead to effective changes of distribution shapes.

Rate-changing nuisances introduced in the subsequent sections are modeled through log-normal distributions whose widths are based on a-priori knowledge from previous measurements or on theoretical reasoning (cf. Sect. 5.3.1). Prior distributions of nuisances with an additional shape-changing effect are constructed by a template morphing algorithm that incorporates the nominally simulated shape of a variable distribution in conjunction with two variants resulting from variations of the respective uncertainty (cf. Sect. 5.3.1). In the following, all shape-changing nuisances also imply rate-changing effects if not stated otherwise.

The set of experimental, theoretical, and modeling uncertainties is discussed below. Rate- and shape-changing effects on the DNN discriminant distributions are representatively shown for the $t\bar{t}H$ discriminant in the "SL, $\geq$6j, $t\bar{t}H$" category, i.e., events with at least six jets in the single-lepton channel that received the highest DNN output in the $t\bar{t}H$ unit. This section concludes with a tabular summary of the statistical model.

8.1.1 Experimental Uncertainties

Sources of uncertainties on experimentally determined quantities are related to limited measurement accuracies. This analysis considers uncertainties regarding the measurement of the integrated luminosity, the calibration of the jet energy scale and resolution, corrections applied to event weights (cf. Sect. 6.4), and the limited amount of simulated events that affect the precision of predicted distributions. They are fully correlated among all processes and explained in the following paragraphs. Luminosity and event correction uncertainties are incorporated through variations of their contribution to the overall event weight (Eqs. 6.1 and 6.7).

Integrated Luminosity

The measurement of the integrated luminosity for the data-taking period in 2016 was performed using the Pixel Cluster Counting (PCC) method (cf. Sect. 3.2.5). The relative uncertainty is estimated from data to be $\pm2.5\%$ [1]. It is incorporated as a single nuisance parameter that equally affects the expected rates of all involved physics processes. The effect on the DNN discriminant in the "SL, $\geq$6j, $t\bar{t}H$" category is visualized in Fig. 8.1a.

Pileup

The amount of additional proton-proton interactions within the same bunch crossing is corrected in simulations by means of event weights to match the recorded pileup

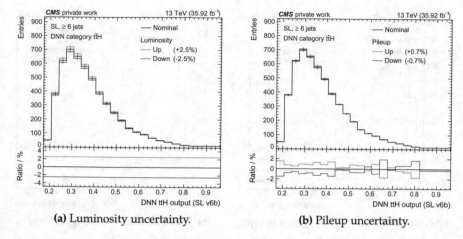

(a) Luminosity uncertainty.   **(b)** Pileup uncertainty.

Fig. 8.1 Effect of uncertainty variations on the DNN discriminant in the "SL, $\geq$6j, $t\bar{t}H$" category. Contributions of all expected backgrounds are added. The bottom ratio plot shows relative shape variations. Changes of the total expected yield are quoted for the respective variation

multiplicity distribution in data (cf. Sect. 6.4). Event weights are defined as the ratio of the number of recorded to simulated pileup events. The simulated pileup profile is parametrized by the total inelastic proton-proton cross section $\sigma_{inel.}$, which is estimated to have a relative uncertainty of $\pm 4.6\%$ [2]. Variations of $\sigma_{inel.}$ are propagated to changes of event weights as shown in Fig. 6.2. Since the value of a weight is a function of the number of pileup events, the associated nuisance parameter has a shape-changing effect, which is demonstrated in Fig. 8.1b. Despite the normalization of the distribution of pileup weights to unity, the rate of predicted contributions can change due to a slightly different pileup profile in the phase space of the particular category. However, the effect only amounts to a change of the predicted yield of $\pm 0.7\%$ in the "SL, $\geq$6j, $t\bar{t}H$" category. The uncertainty is treated as correlated between all processes.

Lepton Efficiencies

Differences between lepton identification and reconstruction efficiencies measured in simulation and data are corrected by deriving appropriate scale factors that are incorporated as event weights (cf. Sect. 6.4). Efficiencies are determined using tag-and-probe methods in dedicated analyses of $Z/\gamma^* \rightarrow e^+e^-/\mu^+\mu^-$ events for different stages of the lepton measurement procedure [3, 4].

For the electron identification efficiency, the number of probe electrons passing the identification criteria is estimated from a fit to the dielectron mass spectrum around the Z boson mass peak [3]. The dominant source of uncertainty is related to the functional model used in the fit. Thus, the efficiency uncertainty is obtained by variations of the fit function. Differences among different event generators are found to be negligible [3]. The same procedure is applied for estimating the uncertainty of the electron reconstruction efficiency. As the probe electron is required to be isolated,

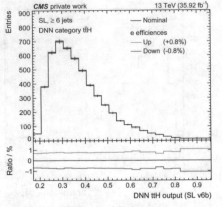
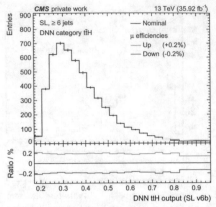

(a) Combined electron efficiency uncertainty.   **(b)** Combined muon efficiency uncertainty.

Fig. 8.2 Effect of uncertainty variations on the DNN discriminant in the "SL, $\geq$6j, $t\bar{t}H$" category. Contributions of all expected backgrounds are added. The bottom ratio plot shows relative shape variations. Changes of the total expected yield are quoted for the respective variation

which is related to a series of kinematic selection criteria (cf. Sect. 6.3.2), variations of these criteria constitute the second highest source of systematic uncertainty [3]. As the overall uncertainty on electron efficiencies is expected to be in the order of ~1%, the different sources are conservatively treated as fully correlated. They are modeled via a nuisance parameter with rate- and shape-changing effect, whose influence on the DNN discriminant in the "SL, $\geq$6j, $t\bar{t}H$" category is shown in Fig. 8.2a.

A similar procedure is employed for the estimation of muon efficiency uncertainties. The number of probe muons is extracted by a parametrized fit of the dimuon mass spectrum to the Z boson mass peak [4]. The main source of uncertainty arises from the choice of the fit model and is estimated by using alternative functions. Moreover, transverse momentum and isolation criteria of the tag muon selection are altered to estimate uncertainties resulting from tag-and-probe conditions, such as the contamination of the measurement phase space with background events [4]. Muon identification, isolation, and tracking efficiencies are conservatively treated as correlated and incorporated as a single nuisance parameter in the statistical model. Its influence on the shape of the discriminants is presented in Fig. 8.2b.

Trigger Efficiencies

Efficiencies of the deployed trigger path configurations (cf. Sect. 6.3.3) are measured using tag-and-probe methods with $Z/\gamma^* \rightarrow e^+e^-/\mu^+\mu^-$ events in the single-lepton channel [3–6], and additionally with $t\bar{t} \rightarrow b\bar{b}\nu\bar{\nu}'l^+l'^-$ events in the dilepton channel [7–9]. Scale factors are applied to correct for discrepancies observed between simulation and data, and are translated to event weights accordingly (cf. Sect. 6.4). They are depicted in Figs. 6.3 and 6.4.

Uncertainties on single-lepton trigger scale factors are estimated analogously to those for lepton efficiencies as described above. The uncertainty for the muon trigger amounts to ~0.5% [4], and to ~2 − 4% for the electron trigger, depending on the

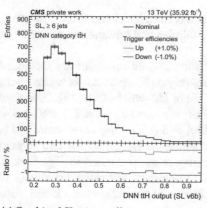 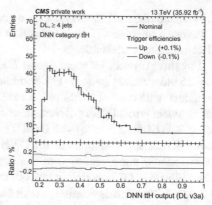

(a) Combined SL trigger efficiency uncertainty.   **(b)** Combined DL trigger efficiency uncertainty.

Fig. 8.3 Effect of uncertainty variations on the DNN discriminant in the "SL, $\geq$6j, $t\bar{t}H$" category (**a**) and in the "DL, 4j, $t\bar{t}H$" category (**b**). Contributions of all expected backgrounds are added. The bottom ratio plot shows relative shape variations. Changes of the total expected yield are quoted for the respective variation

transverse momentum and supercluster pseudorapidity of the probe electron [5]. One nuisance with shape-changing effect is introduced per trigger scale factor. As shown in Fig. 8.3a, their combined impact on the DNN discriminant shapes is relatively small and leads to a change of the expected event yield in the "SL, $\geq$6j, $t\bar{t}H$" category of $\sim$1%.

As explained in Sect. 6.4, scale factors for triggers in the dilepton channel are inferred by monitoring the rates of a weakly correlated cross trigger that involves selection criteria on the amount of missing transverse energy. Uncertainties on the scale factors are estimated based on various considerations. Efficiencies are compared between simulated Drell-Yan and $t\bar{t}$ samples, the missing transverse energy requirement of the cross trigger is varied, and the effect of further selection criteria is studied. Utilized event samples are split according to jet and primary vertex multiplicities and the estimation is repeated per category. The total uncertainty is constructed conservatively using Clopper-Pearson intervals [10] and ranges from $\sim$0.1% in the central, to $\sim$1% in the outer detector regions (Fig. 6.4). Owing to similarities of the measurement methodology, uncertainties between the three dilepton channels are considered as fully correlated and thus modeled by a single nuisance parameter in the statistical model. As shown in Fig. 8.3b, the shape- and rate-changing effects on the discriminant in the "DL, 4j, $t\bar{t}H$" category are almost negligible.

Jet Energy Scale and Resolution

As explained in Sect. 3.3.4, the energy scale and resolution of reconstructed jets are calibrated in a manifold correction procedure, which is explained in detail in Ref. [11]. Owing to the high jet multiplicity in the investigated $t\bar{t}H$ ($H \to b\bar{b}$) final state, this analysis is potentially sensitive to the applied jet corrections and their corresponding uncertainties. Furthermore, a multitude of variables that are considered as inputs to the utilized deep neural networks are constructed from jet observables. Therefore, a thorough treatment of both correction uncertainties and correlations to the DNN discriminants is advisable.

For this purpose, jet energy scale (JES) calibration uncertainties are factorized into 26 independent sources [11]. Some sources rely on comparisons between events generated with either PYTHIA (v6 or v8) [12, 13] or HERWIG++ (v2) [14]. Others are extracted individually for different pseudorapidity regions, i.e., $|\eta| < 1.3$ (BB), $1.3 < |\eta| < 2.5$ (EC1), $2.5 < |\eta| < 3.0$ (EC2), and $|\eta| > 3$ (HF). They are summarized in Table 8.1. Shape variations of the DNN discriminants caused by the two sources of systematic uncertainties with the largest impact are presented in Fig. 8.4a, b. The "FlavorQCD" uncertainty induces the most dominant normalization change of $+2.6/-4.8\%$. This is consistent with the objective of the multi-class classification

Table 8.1 Summary of systematic uncertainties related to jet energy scale corrections [11]

Source name	Description
Absolute corrections (7)	
AbsoluteScale	Constant scale correction uncertainty in global calibration fit
AbsoluteStat	Residual component of statistical uncertainty in global calibration fit
AbsoluteMPFBias	Absolute uncertainties on ISR/FSR corrections
SinglePionECAL	Fitted energy scale in the electromagnetic calorimeter
SinglePionHCAL	Fitted energy scale in the hadronic calorimeter
Fragmentation	Generator difference between jet fragmentation in high-p_T regime
TimePtEta	Time dependence of jet energy corrections on the data-taking period
Relative corrections (12)	
RelativeBal	Difference in calibration fits between p_T-balance and MPF[a] methods
RelativeFSR	Uncertainty due to η-dependence of ISR/FSR corrections
RelativeJEREC1/EC2/HF	Uncertainty due to η-dependence of jet energy resolution scale factors
RelativePtBB/EC1/EC2/HF	Difference between log-linear and constant p_T calibration fits
RelativeStatEC/HF/FSR	Statistical uncertainty of the η-dependence determination
Pileup offset corrections (6)	
PileUpDataMC	Variation of the pileup offset correction scale factor by 5%
PileUpPtBB/EC1/EC2/HF	Uncertainty due to p_T-dependence of pileup offset correction
PileUpPtRef	Uncertainty due to p_T-dependence of pileup offset correction in data
Jet flavor corrections (1)	
FlavorQCD	Uncertainties on flavor-dependent jet responses in studied dijet events

[a]MPF refers to missing transverse momentum projection fraction methods [15]

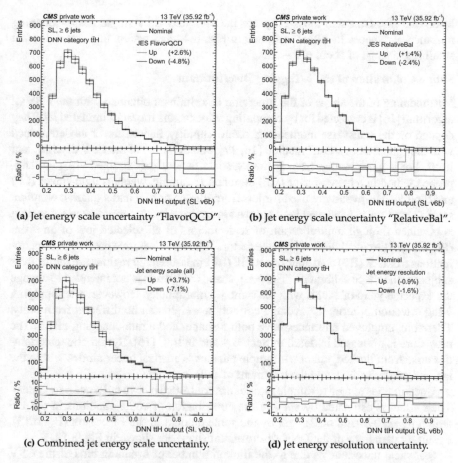

(a) Jet energy scale uncertainty "FlavorQCD".

(b) Jet energy scale uncertainty "RelativeBal".

(c) Combined jet energy scale uncertainty.

(d) Jet energy resolution uncertainty.

Fig. 8.4 Effect of uncertainty variations on the DNN discriminant in the "SL, $\geq$6j, $t\bar{t}H$" category. Contributions of all expected backgrounds are added. The bottom ratio plot shows relative shape variations. Changes of the total expected yield are quoted for the respective variation

strategy pursued in this analysis since the trained deep neural networks are sensitive to variables that are correlated with jet-flavor information. The combined effect of all uncertainties on the discriminant shape is shown in Fig. 8.4c. In total, 26 uncorrelated nuisance parameters are included in the statistical model with shape- and rate-changing effects.

The jet energy and momentum resolution is found to be superior in simulated events with respect to data. Scale factors are extracted on a per-jet basis that correct for differences between transverse momenta of reconstructed jets and geometrically matched jets on generator level. These scale factors are calibrated using dijet and $\gamma +$ jets data events by exploiting their momentum balance in the transverse plane. The method is described in Sect. 6.3.5 and more detailed in Ref. [11]. Uncertainties result from initial- and final-state corrections, imbalance of the dijet system on particle-level, and non-Gaussian tails. They are incorporated via a single shape-changing nuisance parameter. Its effect is shown in Fig. 8.4d and results in a one-sided but small rate change of about $\sim 1 - 2\%$.

Shape Calibration of the *b*-Tagging Discriminant

Mismodeling of the shape of the *b*-tagging discriminant obtained with the CSVv2 algorithm [16] is corrected for by computing scale factors for each simulated jet. They depend on the transverse momentum, pseudorapidity, hadron flavor as determined with the "ghost association" method [16, 17], and the value of the CSV discriminant itself. For jets originating from *b* quarks, scale factors are extracted via a tag-and-probe method in a heavy-flavor (HF) enriched $t\bar{t}$ sample, whereas light-flavor (LF) enriched *Z*+jets events are used for jets from gluons or *u*, *d* and *s* quarks. Nominal scale factors for jets induced by *c* quarks are set to unity. A corrective event weight is obtained through multiplication of scale factors of all selected jets of an event (Eq. 6.11). The calculation of jet scale factors follows an iterative method as contaminations from LF (HF) contributions in HF (LF) measurement regions are themselves subject to the shape calibration. The *b*-tag shape calibration is not intended to change the expected yield of events with the same jet multiplicity. However, after applying *b*-tag selection criteria, the average correction weight can be different from unity. Therefore, employed nuisances have both a shape- and a rate-changing effect. The procedure is discussed in detail in Sect. 6.4 and in Ref. [16]. Systematic uncertainties arise from limited, unknown sample purities, statistical uncertainties within the tag-and-probe method, and the treatment of *c*-flavored jets.

Uncertainties related to sample purities are controlled by two nuisance parameters. Their effect is estimated by varying the contribution of contaminations in the HF region by $\pm 20\%$ (Eq. 6.12), and by the same amount in the LF region (Eq. 6.13). Resulting variations of the DNN discriminant shape are shown in Fig. 8.5a, b.

Statistical uncertainties due to the limited number of events in bins of the CSV discriminant are modeled by two independent envelopes. They are constructed by varying the content of each bin by its statistical uncertainty, multiplied by either a linear function to model a possible tilt of the discriminator distribution, or a quadratic function to cover fluctuations of the data-to-simulation ratio in the center of the distribution. This prescription is applied separately in the HF and in the LF region, leading to four independent nuisance parameters.

Uncertainties for *c*-flavored jets are based on those for *b*-flavored jets. The two terms for statistical and purity uncertainties are conservatively doubled and added in quadrature. Subsequently, the resulting relative uncertainty is used to construct two independent envelopes following the approach explained above, which are modeled by means of two additional nuisance parameters. The impact on the DNN discriminant shape of the linear component is depicted in Fig. 8.5c and leads to slightly asymmetric rate changes of $-9.7/+12.9\%$.

The combined effect of all eight nuisances on the DNN discriminant shape, taking relative signs of changes conservatively into account, is shown in Fig. 8.5d. With a total variation of $+17.7/-14.7\%$ on the expected yield, uncertainties related to the *b*-tagging discriminant calibration are expected to constitute the dominant sources of experimental uncertainties.

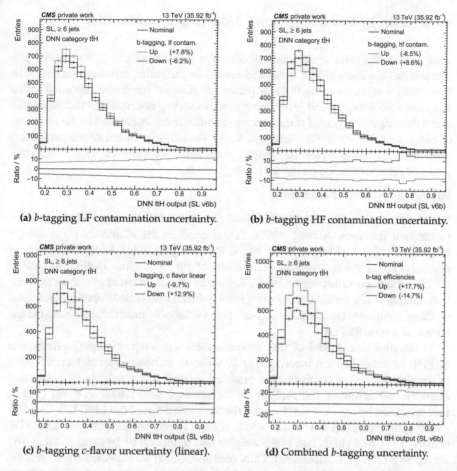

(a) *b*-tagging LF contamination uncertainty.

(b) *b*-tagging HF contamination uncertainty.

(c) *b*-tagging *c*-flavor uncertainty (linear).

(d) Combined *b*-tagging uncertainty.

Fig. 8.5 Effect of uncertainty variations on the DNN discriminant in the "SL, $\geq$6j, $t\bar{t}H$" category. Contributions of all expected backgrounds are added. The bottom ratio plot shows relative shape variations. Changes of the total expected yield are quoted for the respective variation

Uncertainty Due to Limited Amount of Simulated Events

The treatment of statistical uncertainties due to the limited amount of simulated events in bins of the DNN discriminants is described in Sect. 5.3.1. Per discriminant bin, a rate-changing nuisance parameter is introduced to the statistical model, which is correlated among all expected physics process contributions in the particular bin. The width of the corresponding a-priori distribution is estimated from the Poisson uncertainty of the sum of simulated events. The binning optimization algorithm discussed in Sect. 7.3 ensures that a minimum amount of events is contained in each discriminant bin, so that a single Gaussian distribution per bin is sufficient to model statistical uncertainties with reasonable accuracy [18].

8.1.2 Theory and Background Modeling Uncertainties

This section discusses systematic uncertainties that are related to theoretical predictions and the modeling of simulated events. In particular, this analysis considers
uncertainties related to variations of parton distribution functions, renormalization
and factorization scales, and generator-specific modeling uncertainties that especially
affect the shape of expected $t\bar{t}$ background contributions. In contrast to the treatment
of experimental uncertainties (cf. Sect. 8.1.1), the majority of nuisances parameters
introduced in the subsequent paragraphs affect either a specific physics process or a
group of processes based on theoretical arguments.

Parton Distribution Functions

Parton distribution functions (PDFs) provide a probabilistic description of parton
momentum fractions. Among others, they depend on the momentum transfer Q^2
of the interaction and on the flavor of the involved parton (cf. Sect. 2.2.1). Sets of
multiple PDFs are employed for calculations of $pp \rightarrow X$ cross sections as well as
for the simulation of proton-proton collisions by means of event generators. Variations among PDFs within a set or even among different sets developed by dedicated
working groups are typically used to estimate systematic uncertainties related to the
choice of a specific PDF.

In the statistical model of this analysis, rate- and shape-changing effects due
to PDF uncertainties are incorporated as separate nuisance parameters. Changes
of simulated distribution shapes are modeled through a reweighting of events that
considers variations of 100 replicas of the nominal NNPDF3.0 PDF set [19] as
recommended by the PDF4LHC group for the second run of the LHC [20, 21]. The
reweighting method is described in Ref. [22], is implemented using the LHAPDF
package [23], and yields two event weights that reflect up $(+1\sigma)$ and down (-1σ)
variations of the PDF uncertainty. Only contributions from expected signal and the
dominant $t\bar{t}$ backgrounds are considered. In order to conserve nominally predicted
event yields, weights are normalized per simulated sample such that their averages are
equal to unity before applying any selection criteria. Therefore, slight rate-changes
can emerge after the event selection. The shape-changing effect of the corresponding
nuisance parameter is shown in Fig. 8.6a.

PDF uncertainties related to fixed-order cross section calculations (Table 2.4) are
described by four nuisance parameters that affect the normalization of one process, or
correlated between multiple processes depending on their leading production mechanism in proton-proton collision events. A nuisance parameter for gg-induced events
accounts for 4% uncertainty on $t\bar{t}$, and for 3% on $t\bar{t}+Z$ events. The uncertainty on
single $t/\bar{t}$ production from qg interactions amounts to 3%. A third nuisance parameter for $q\bar{q}$-induced processes covers an uncertainty of 4% on V+jets events, and
of 2% on both diboson and $t\bar{t}+W$ production. Lastly, a separate nuisance parameter accounts for an expected PDF uncertainty on $t\bar{t}H$ production of 3.6%. In the
"SL, $\geq$6j, $t\bar{t}H$" category, the combined cross section uncertainty amounts to relative
changes of the expected yield of $\pm3.9\%$ as shown in Fig. 8.6b.

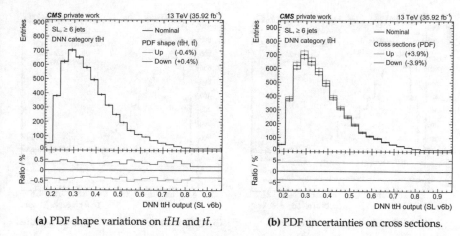

(a) PDF shape variations on $t\bar{t}H$ and $t\bar{t}$.   (b) PDF uncertainties on cross sections.

Fig. 8.6 Effect of uncertainty variations on the DNN discriminant in the "SL, $\geq$6j, $t\bar{t}H$" category. Contributions of all expected backgrounds are added. The bottom ratio plot shows relative shape variations. Changes of the total expected yield are quoted for the respective variation

Renormalization and Factorization Scales

Renormalization and factorization scales define the value of the cutoff scale for absorbing infrared and ultraviolet divergences, respectively, in the calculation of scattering amplitudes (cf. Sect. 2.2.1). A coherent scale $\mu_R = \mu_F$ is selected according to the momentum scale of the hard scattering process, such as m_t for processes involving top quarks. Related systematic uncertainties are typically estimated through independent variations of μ_R and μ_F by factors of $\frac{1}{2}$ and 2. They are propagated to uncertainties of cross section calculations and to variations of distributions simulated with event generators. Similar to the treatment of PDF uncertainties as discussed above, shape- and rate-changing effects are considered individually by means of multiple nuisance parameters.

Shape variations of simulated distributions are modeled through a reweighting technique [24]. The probability to obtain a specific event is determined for three independent renormalization and factorization scales of m_t (nominal), $2m_t$ ($+1\sigma$), and $m_t/2$ (-1σ). The ratios between the nominal and varied probabilities serve as event weights. Average weights are normalized to unity prior to the application of selection criteria, potentially resulting in minor rate-changes after the event selection. Two independent nuisance parameters with shape-changing effect are included in the statistical model and affect the dominant $t\bar{t}$ background contribution. Their combined impact on the DNN discriminant shape is presented in Fig. 8.7a.

Variations of renormalization and factorization scales are reflected in cross section uncertainties (Table 2.4) through five nuisance parameters with rate-changing effect. Uncertainties of $+2/-4\%$ on $t\bar{t}$, $+13/-12\%$ on $t\bar{t}+W$, and $+10/-12\%$ on $t\bar{t}+Z$ processes are treated as correlated through a single nuisance parameter. Three uncorrelated parameters are included for uncertainties of $+3/-2\%$ on single $t/\bar{t}$, $\pm1\%$ on V+jets, and $\pm2\%$ on diboson production, respectively. The

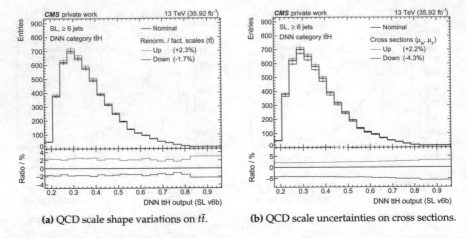

(a) QCD scale shape variations on $t\bar{t}$.   **(b)** QCD scale uncertainties on cross sections.

Fig. 8.7 Effect of uncertainty variations on the DNN discriminant in the "SL, $\geq$6j, $t\bar{t}H$" category. Contributions of all expected backgrounds are added. The bottom ratio plot shows relative shape variations. Changes of the total expected yield are quoted for the respective variation

expected scale uncertainty on $t\bar{t}H$ production of $+5.8/-9.2\%$ is covered by a separate nuisance parameter. As shown in Fig. 8.7b, the combined change of the expected yield in the "SL, $\geq$6j, $t\bar{t}H$" category amounts to $+2.2/-4.3\%$.

Modeling of $t\bar{t}$ Contributions

Contributions from $t\bar{t}$ processes constitute the main background to the search for $t\bar{t}H$ ($H \rightarrow b\bar{b}$) production (cf. Sect. 4.1). Therefore, an accurate modeling of simulated variable distributions and a reliable description of systematic uncertainties are of particular importance. Corresponding generator tuning parameters are summarized by the CUETP8M2T4 tune [25, 26]. In addition to uncertainties related to parton density functions as well as to QCD scales in fixed-order scattering amplitude calculations as discussed above, parton showering and the modeling of global hadron production, also referred to as the "underlying event" (UE), are considered for simulations of $t\bar{t}$ events. Scale uncertainties in the modeling of parton showers are treated separately for initial- (ISR) and final-state radiation (FSR) processes. The nominal scales are varied by factors of 2 and $\frac{1}{2}$. Matching between parton shower and matrix element calculations (ME-PS) is controlled through a damping parameter h_{damp} for high-p_T radiations from POWHEG, whose value is determined as $1.58^{+0.66}_{-0.59}\, m_t$ for analyses in the phase space of top quark pair production [27, 28]. Parameters for the tune of the underlying event description and associated uncertainty variations are described in Ref. [25].

Separate $t\bar{t}$ event samples were generated for ISR, FSR, ME-PS, and UE variations, and applied to identical selection and classification procedures. The amount of available statistics, however, is found to be insufficient to properly model shape variations of the DNN discriminant distributions. Therefore, systematic uncertainties are estimated conservatively by adding in quadrature the statistical uncer-

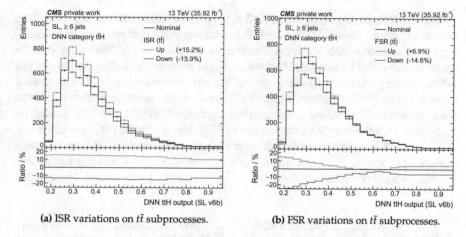

(a) ISR variations on $t\bar{t}$ subprocesses.   (b) FSR variations on $t\bar{t}$ subprocesses.

Fig. 8.8 Effect of uncertainty variations on the DNN discriminant in the "SL, $\geq$6j, $t\bar{t}H$" category. Contributions of all expected backgrounds are added. The bottom ratio plot shows relative shape variations. Changes of the total expected yield are quoted for the respective variation

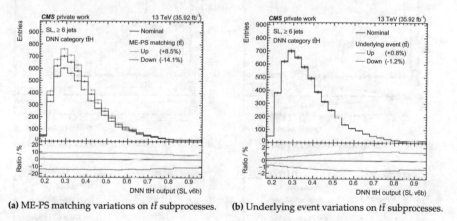

(a) ME-PS matching variations on $t\bar{t}$ subprocesses.   (b) Underlying event variations on $t\bar{t}$ subprocesses.

Fig. 8.9 Effect of uncertainty variations on the DNN discriminant in the "SL, $\geq$6j, $t\bar{t}H$" category. Contributions of all expected backgrounds are added. The bottom ratio plot shows relative shape variations. Changes of the total expected yield are quoted for the respective variation

tainty in each bin of the varied DNN discriminants. The difference between the resulting template and the nominal discriminant distribution is incorporated as a rate-changing nuisance. This procedure is done individually for each of the five $t\bar{t}$ subprocesses, i.e., $t\bar{t}+b\bar{b}$, $t\bar{t}+b/\bar{b}$, $t\bar{t}+2b$, $t\bar{t}+c\bar{c}$, and $t\bar{t}+$lf, resulting in a total of 20 nuisances parameters. The effect on the discriminant in the "SL, $\geq$6j, $t\bar{t}H$" category is presented in Fig. 8.8a, b for ISR and FSR, and in Fig. 8.9a, b for ME-PS and UE variations, respectively. With the exception of UE uncertainties, they cause differences of the expected event yield of about $\pm15\%$.

Furthermore, as discussed in Sect. 2.2.4, the production of additional b jets can be caused by various topologies in association with top quark pair production. Depending on the scale of additional heavy-flavor emissions, initial- and final-state $g \to b\bar{b}$ splittings can either be described by hard $t\bar{t}$ matrix elements including parton-shower branchings, by collinear splittings during parton showering, or by $t\bar{t}+b\bar{b}$ matrix elements [29]. Neither higher-order calculations nor experimental methods can currently constrain expected rates with an accuracy better than 35% [29–32]. Therefore, additional, conservative rate uncertainties of 50% are considered for $t\bar{t}+b\bar{b}$, $t\bar{t}+b/\bar{b}$, $t\bar{t}+2b$, and $t\bar{t}+c\bar{c}$ contributions, which are incorporated through four uncorrelated nuisance parameters. Figure 8.10a–d show their effect on the DNN discriminant in the "SL, $\geq$6j, $t\bar{t}H$" category. Rate uncertainties relative to the overall background

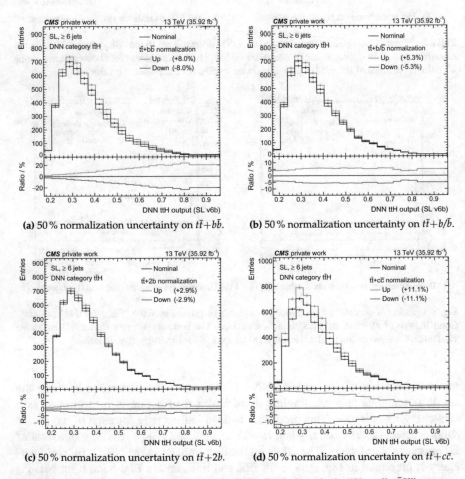

(a) 50 % normalization uncertainty on $t\bar{t}+b\bar{b}$.

(b) 50 % normalization uncertainty on $t\bar{t}+b/\bar{b}$.

(c) 50 % normalization uncertainty on $t\bar{t}+2b$.

(d) 50 % normalization uncertainty on $t\bar{t}+c\bar{c}$.

Fig. 8.10 Effect of uncertainty variations on the DNN discriminant in the "SL, $\geq$6j, $t\bar{t}H$" category. Contributions of all expected backgrounds are added. The bottom ratio plot shows relative shape variations. Changes of the total expected yield are quoted for the respective variation

expectation consequently increase in bins with high fractions of events of the particular $t\bar{t}$ subprocess, as visible in the presented ratio diagrams. Owing to the signal-like final state of $t\bar{t}+b\bar{b}$ events, dominant uncertainties emerge in bins with high discriminant values (Fig. 8.10a), whereas those related to more distinguishable $t\bar{t}+c\bar{c}$ events

Table 8.2 Summary of systematic uncertainties considered in the statistical model of this analysis, listing uncertainty sources, whether associated nuisance parameters affect only rates or also shapes of DNN discriminant distributions, the number of nuisance parameters N_p, and additional remarks

Uncertainty source	Shape	N_p	Remarks
EXPERIMENTAL			
Integrated luminosity		1	2.5% per process
Pileup	✓	1	From 4.6% $\sigma_{\text{inel.}}$ uncertainty
Trigger efficiencies	✓	5	Per lepton channel
Electron efficiencies	✓	1	ID, reconstruction
Muon efficiencies	✓	1	ID, isolation, tracking
Jet energy resolution	✓	1	
Jet energy scale	✓	26	Factorized sources
b-tagging			
Heavy-flavor contamination	✓	1	Varied by 20% in SF calculation
Light-flavor contamination	✓	1	Varied by 20% in SF calculation
Heavy-flavor statistics	✓	2	Linear and quadratic terms
Light-flavor statistics	✓	2	Linear and quadratic terms
c-flavor uncertainty	✓	2	Linear and quadratic terms
Amount of simulated events	✓	455	1 per discriminant bin
THEORETICAL			
Parton distribution functions			
NNPDF3.0 set replicas ($t\bar{t}H$, $t\bar{t}$)	✓	1	Variation from 100 replicas
Cross section (gg, $t\bar{t}H$)		1	Affects $t\bar{t}H$
Cross sections (gg)		1	Affects $t\bar{t}$, $t\bar{t}+Z$
Cross sections (qg)		1	Affects single $t\bar{t}$
Cross sections ($q\bar{q}$)		1	Affects V+jets, $t\bar{t}+W$, diboson
Renorm./fact. scales			
Renormalization scale ($t\bar{t}$)	✓	1	μ_R varied by 2 and 1/2
Factorization scale ($t\bar{t}$)	✓	1	μ_F varied by 2 and 1/2
Cross section ($t\bar{t}H$)		1	NLO scale uncertainty
Cross section ($t\bar{t}$, $t\bar{t}+V$)		1	NNLO/NLO scale uncertainty
Cross section (single $t\bar{t}$)		1	NLO scale uncertainty
Cross section (V+jets)		1	NNLO scale uncertainty
Cross section (diboson)		1	NLO scale uncertainty
$t\bar{t}$ modeling			
ME-PS matching		5	Per $t\bar{t}$ process
Initial-state radiation		5	Per $t\bar{t}$ process
Final-state radiation		5	Per $t\bar{t}$ process
Underlying event		5	Per $t\bar{t}$ process
$t\bar{t}$+hf normalization		4	50% per $t\bar{t}$+hf process

are found at low discriminant values. Therefore, the nuisance parameter accounting for the normalization of $t\bar{t}+b\bar{b}$ contributions is expected to cause a significant impact in the statistical evaluation.

8.1.3 Summary of Systematic Uncertainties

A summary of all systematic uncertainties considered in this analysis is presented in Table 8.2. They are grouped into sources of either experimental or theoretical origin. The table also quotes the number of corresponding nuisance parameters in the statistical model and whether they involve a purely rate-changing effect on one or more physics processes or also affect the shapes of DNN discriminant distributions that are used in the signal extraction procedure. Except for the nuisances that account for limited simulation statistics per discriminant bin, the uncertainty model contains a total of 80 nuisance parameters, of which 46 entail a shape-changing effect.

8.2 Measurement Results

This section presents and discusses the results of the signal extraction procedure by means of the statistical inference methods introduced in Sect. 5.3. The procedure relies on the distributions of DNN discriminants in categories defined by the number of leptons, the number of jets, and the most probable process as assigned by the DNN multi-class classification approach (cf. Sect. 7). Results are extracted through a binned maximum likelihood fit to data given the statistical model explained in the previous section, and quoted in terms of expected and observed upper limits at a 95% confidence level on the signal strength modifier $\mu = \sigma/\sigma_{SM}$, its best fit value and uncertainty, and the corresponding exclusion significance under the background-only hypothesis. The section concludes with an evaluation of the influence of individual sources of systematic uncertainties.

8.2.1 Expected and Observed Limits

Upper limits on the signal strength modifier μ are set at a 95% confidence level following a modified frequentist CL_S technique with the asymptotic prescription [33–35] (cf. Sect. 5.3). Expected upper limits are evaluated for a statistical model that assumes $\mu = 0$, so that values below one signify incompatibility of the background-only hypothesis given a simulated dataset that includes signal according to the SM prediction.

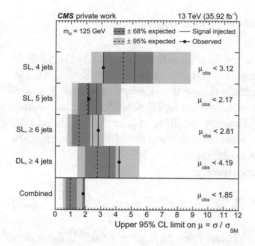

Category	Expected	Observed
SL, 4 jets	$4.41^{+1.91}_{-1.27}$	3.12
SL, 5 jets	$2.09^{+0.93}_{-0.62}$	2.17
SL, ≥ 6 jets	$1.55^{+0.70}_{-0.46}$	2.81
DL, ≥ 4 jets	$2.76^{+1.18}_{-0.79}$	4.19
Combined	$0.99^{+0.42}_{-0.29}$	1.85

Fig. 8.11 Observed (black markers) and median expected (dashed line) upper limits at 95% confidence on the signal strength modifier $\mu = \sigma/\sigma_{SM}$ per analysis category and combined for all categories. Uncertainty bands denote central 68% (green) and 95% (yellow) intervals of the distribution of limits obtained for the expected background-only hypothesis. Injected limits (red line) represent the median expected limit under the signal hypothesis with $\mu = 1$

Obtained upper limits are shown in Fig. 8.11 separately for the four categories defined by the lepton channel and the jet multiplicity, and for their combination. Median expected limits are marked by dashed lines and their corresponding central 68% and 95% intervals are denoted by green and yellow uncertainty bands, respectively. The median expected limits for the signal hypothesis are depicted by red lines. Black markers refer to the observed upper limits given the measured data.

The combined expected upper limit is measured as $0.99^{+0.42}_{-0.29}$. This finding suggests that the sensitivity provided by the presented analysis is sufficient to disfavor the background-only hypothesis in case of a $t\bar{t}H$ cross section as predicted by the SM with more than 95% confidence. With upper limits of $2.09^{+0.93}_{-0.62}$ and $1.55^{+0.70}_{-0.46}$, single-lepton events with five and at least six jets, respectively, exhibit the largest influence on the measurement, underlining the expectation based on predicted signal-to-background ratios in these categories (Tables 7.6 and 7.7).

The combined observed upper limit is measured as 1.85, which is compatible with the SM prediction. More precisely, the cross section of $t\bar{t}H$ production is below $1.85 \times \sigma_{SM}$ with 95% confidence. The obtained value is also in agreement with the expected upper limit under the signal hypothesis (red line). When focusing on the individual categories, this comparison yields statistical under-fluctuations in data in the single-lepton categories with four or five jets, and slight over-fluctuations in the two remaining categories. Considering the relatively large 68% and 95% uncertainty intervals, however, this finding appears statistically plausible.

8.2.2 $t\bar{t}H$ Production Cross Section and Significance

The signal strength modifier μ is extracted in a binned maximum likelihood fit to data, simultaneously over all bins of the 24 DNN discriminants (cf. Sect. 5.3). The parameter is allowed to float unconstrainedly during the fit, whereas nuisance parameters θ model the effect of systematic uncertainties under consideration of suitable a-priori distribution functions. In the following, resulting maximum likelihood estimators are referred to as best fit values. Moreover, the best fit value for μ is obtained via profiling of nuisance parameters as explained in Sect. 5.3.2.

The observed best fit value μ_{obs} is measured as

$$\mu_{\text{obs}} = 0.98 \, ^{+0.50}_{-0.48} = 0.98 \, ^{+0.26}_{-0.25} \, (\text{stat.}) \, ^{+0.43}_{-0.41} \, (\text{syst.}), \tag{8.3}$$

which is in good agreement with the expected value of $\mu_{\text{exp}} = 1.00^{+0.55}_{-0.49}$ from simulation. Based on the SM prediction of $\sigma_{SM} = 507^{+35}_{-50}$ fb with NLO QCD accuracy and NLO electroweak corrections [36] (cf. Sect. 2.2.2), this yields a cross section for $t\bar{t}H$ production of

$$\sigma_{t\bar{t}H} = 497 \, ^{+252}_{-243} \, \text{fb} = 497 \, ^{+130}_{-128} \, (\text{stat.}) \, ^{+216}_{-206} \, (\text{syst.}) \, \text{fb}, \tag{8.4}$$

corresponding to a relative uncertainty of $+51/-49\%$. A detailed breakdown of the systematic uncertainty into specific sources is presented in the next section. The observed significance for excluding the background-only hypothesis is measured to be 2.04σ above the expected background and corresponds to a p-value of 2.08×10^{-2}. A significance of 1.92σ with a p-value of 2.72×10^{-2} is expected from simulation. The results are summarized in Table 8.3.

The measured signal strength modifier is compatible with previous searches for $t\bar{t}H$ production in the $H \to b\bar{b}$ channel with single-lepton and dilepton decays of the $t\bar{t}$ system, as conducted by the ATLAS and CMS collaborations and presented in Table 2.5. A comparison between measured cross sections and the SM prediction is shown in Fig. 8.12. With an uncertainty of $+0.50/-0.48$ in units of σ_{SM}, the presented analysis exhibits a competitive accuracy, located between the uncertainty obtained by the ATLAS collaboration of $+0.64/-0.61$ [37], and a value of

Table 8.3 Summary of expected and observed analysis results in terms of the best fit value of the signal strength modifier $\mu = \sigma/\sigma_{\text{SM}}$, the significance in Gaussian standard deviations above the expected background, and the upper limit at 95% CL. The expected upper limit is evaluated under the background-only hypothesis

Result	Expected	Observed
Best fit value μ	$1.00^{+0.55}_{-0.49}$	$0.98^{+0.50}_{-0.48}$
Significance	1.92σ	2.04σ
Upper limit on μ	$0.99^{+0.42}_{-0.29}$	1.85

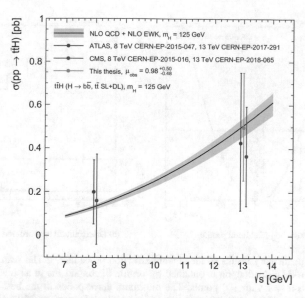

Fig. 8.12 Illustration of the $t\bar{t}H$ cross section as predicted by theoretical calculations [36], measurements performed by the CMS [38, 39] and ATLAS experiments [37, 40], and as measured in this thesis. The measured cross section values emerge from the transformation of the signal strength modifier $\mu = \sigma/\sigma_{SM}$, which does not follow any prior constraint in the fitting procedure. Therefore, in case of statistical under-fluctuations in data, the measured $[+1\sigma, -1\sigma]$ interval might have a negative component, as visible in the CMS measurement at 8 TeV

$| 0.45/-0.45$ as measured by the CMS collaboration [38]. It should be noted that the analysis in the single-lepton channel as presented in this thesis is included in the result of the CMS collaboration [38]. The quoted analysis uses an alternative strategy in the dilepton channel based on a combination of boosted decision trees and a matrix element method discriminant.

A validation of the signal extraction procedure is depicted in Fig. 8.13a. It shows the profile of the negative log-likelihood (NLL) curve whose minimum is determined through a scanning process with 100 points around the SM expectation of $\mu = 1$. The horizontal location of the minimum marks the best fit value, whereas the vertical location is shifted to zero. Thus, the uncertainty on the measurement emerges from intersections of the NLL curve with one. Its shape is parabolic with a pronounced minimum, suggesting an effective and stable fitting procedure. Small fluctuations are statistically insignificant and affect neither the position of the minimum, nor the obtained uncertainties.

Besides the determination of the signal strength modifier, the likelihood maximization yields best fit values for each nuisance parameter of the statistical model as well as an estimate of a-posteriori distribution widths. Corresponding values before and after application of the fitting procedure are referred to as "pre-fit" and "post-fit" values, respectively. The difference between central post- and pre-fit nuisance

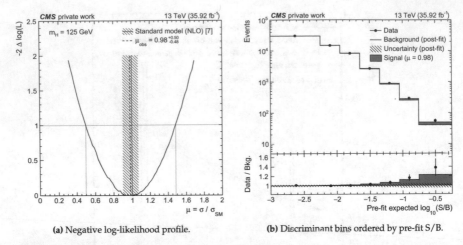

(a) Negative log-likelihood profile.

(b) Discriminant bins ordered by pre-fit S/B.

Fig. 8.13 Supplementary results of the signal extraction procedure. **a** The scan of the negative log-likelihood (NLL) distribution is obtained by profile likelihood fits in an interval around the SM expectation $\mu = 1$ with 100 points. The minimum corresponds to the best fit value and is vertically shifted to zero to expose the fit uncertainty at the intersections of the NLL distribution with one (cf. Sect. 5.3.2). **b** Post-fit event yields contained in bins of the DNN discriminants, ordered by the signal-to-background ratio in each bin according to the pre-fit expectation. The blue area denotes the fitted signal contribution. The shown ratio suggests that, including post-fit uncertainties, the background-only hypothesis is not able to describe recorded data with good agreement. The compatibility is enhanced when including the fitted signal

parameter values, divided by their pre-fit prior distribution widths are called "pulls" (cf. Sect. 5.3).

The agreement between data and simulated events including the $t\bar{t}H$ signal process scaled by the observed best fit value $\mu_{\mathrm{obs}} = 0.98$ is shown in Fig. 8.13b. All bins of the 24 DNN discriminant distributions are ordered by the logarithm of their signal-to-background ratios S/B according to the pre-fit expectation per bin, and their bin contents are inserted into a histogram. The black line denotes post-fit background contributions, i.e., a-priori expected contributions, corrected by nuisance parameter pulls obtained in the fit. The fitted signal contribution is denoted by the blue area. The ratio plot below the histogram suggests that, including post-fit uncertainties, the background-only hypothesis does not describe the recorded data with sufficient agreement. However, better compatibility is achieved when including the signal contribution.

The individual DNN discriminant distributions, after the application of post-fit corrections to shapes and process normalizations, are shown in Figs. 8.14, 8.15, and 8.16 for single-lepton events with four, five, and at least six jets, respectively, and in Fig. 8.17 for dilepton events with at least four jets. The fitted $t\bar{t}H$ signal is included in the stacked contributions of simulated events. Overall, there is very good agreement between data and simulation within statistical uncertainties. Especially for high discriminant values in $t\bar{t}H$-enriched categories, for example, in Figs. 8.16a

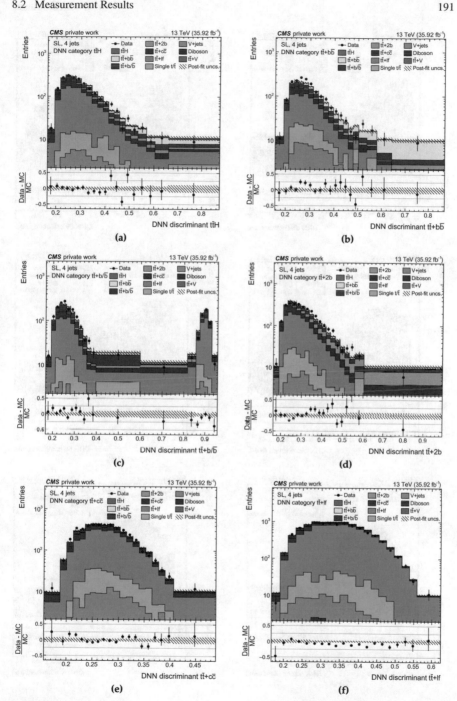

Fig. 8.14 DNN discriminants for single-lepton events with four jets after the fit to data. Hatched areas denote the total uncertainty on expected background contributions and incorporate constraints on uncertainties as obtained in the fit. The first and last bins contain under- and overflow events, respectively

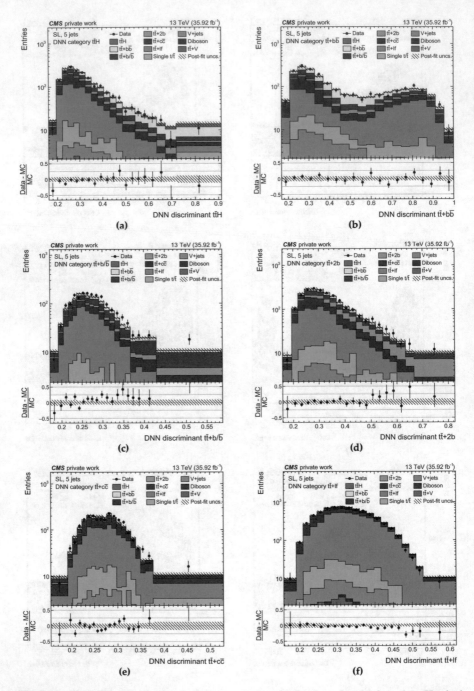

Fig. 8.15 DNN discriminants for single-lepton events with five jets after the fit to data. Hatched areas denote the total uncertainty on expected background contributions and incorporate constraints on uncertainties as obtained in the fit. The first and last bins contain under- and overflow events, respectively

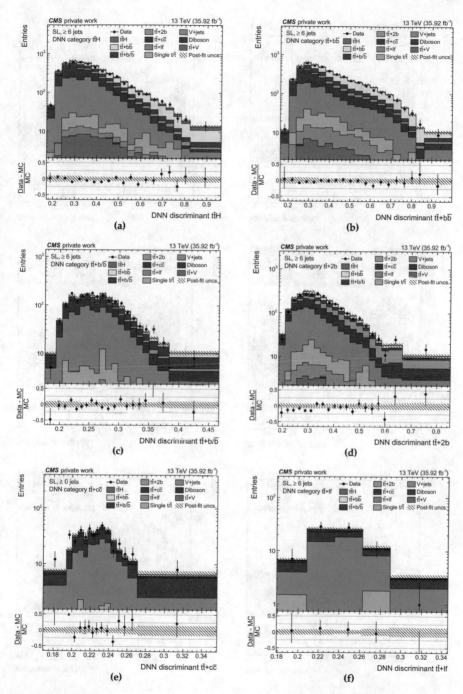

Fig. 8.16 DNN discriminants for single-lepton events with at least six jets after the fit to data. Hatched areas denote the total uncertainty on expected background contributions and incorporate constraints on uncertainties as obtained in the fit. The first and last bins contain under- and overflow events, respectively

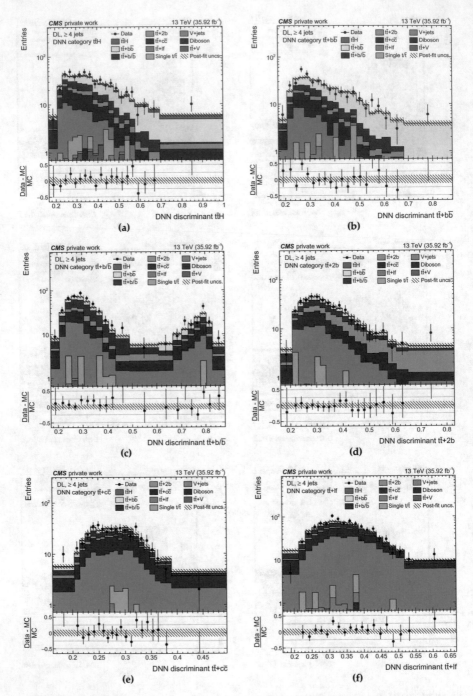

Fig. 8.17 DNN discriminants for dilepton events with at least four jets after the fit to data. Hatched areas denote the total uncertainty on expected background contributions and incorporate constraints on uncertainties as obtained in the fit. The first and last bins contain under- and overflow events, respectively

and 8.15a, the inclusion of the fitted signal appears to improve the agreement with recorded data compared to the background-only scenario. As explained above, this finding is summarized over all discriminant bins in Fig. 8.13b.

The post-fit uncertainties shown in the ratio below each histogram contain all systematic uncertainties, including those due to the limited amount of simulated events. They exhibit reasonable constraints compared to their coverage before the fit (cf. Sect. 7.4). The strengths of constraints on individual nuisance parameters and their impact on the measured results are discussed in the following section.

8.2.3 Impact of Systematic Uncertainties

In this section, the impact of individual sources of systematic uncertainties is studied. First, the total uncertainty in the measurement of the best fit value μ_{obs} is studied through a breakdown into groups of specific uncertainties. Thereafter, a comparison of nuisance parameter values before and after the fit to data is presented alongside their impact on the signal extraction procedure.

The uncertainty in the best fit value μ_{obs} caused by a single source of systematic uncertainty n, modeled by a nuisance parameter θ_n, is estimated as follows. First, the total uncertainty $\Delta\mu_{\text{obs}}$ is obtained by means of the profile likelihood fit as described above. Then, the fit is repeated with the nuisance parameter under study fixed to its post-fit value as determined by the initial fit, resulting in a changed, total uncertainty $\Delta\mu_{\text{obs, fixed } n}$. The contribution of the nuisance to the overall uncertainty, "$\pm\Delta\mu_{\text{obs}, n}$", is then obtained via subtraction in quadrature,

$$\pm \Delta\mu_{\text{obs},n} = \sqrt{(\Delta\mu_{\text{obs}})^2 - (\Delta\mu_{\text{obs, fixed } n})^2}. \tag{8.5}$$

This procedure can also be applied to evaluate the contribution of a group of nuisance parameters. In contrast, the statistical uncertainty is determined by fixing all nuisance parameters to their post-fit values.

A breakdown of uncertainties following this method is shown in Table 8.4. The overall uncertainty is split into its statistical and systematic components. The systematic contribution is further subdivided into theory and background modeling uncertainties, contributions of experimental sources, and the uncertainty due to the size of simulated event samples. The latter is quoted apart from the group of experimental sources owing to their similar contributions. It should be noted that quadratic sums of uncertainties can differ from quoted total values due to dependencies between nuisance parameters within the fitting procedure. For the same reason, sources of systematic uncertainties, which are presumably interdependent by construction of the statistical model, such as $t\bar{t}$+hf normalization and parton shower uncertainties (cf. Sect. 8.1.2), are fixed simultaneously to estimate their contribution. Otherwise, the effect of individual uncertainties might be underestimated by this method.

Table 8.4 Contributions of different groups of systematic uncertainties to the combined, observed uncertainty of the best fit value. The stated differences $\pm\Delta\mu_{obs}$ are obtained by repeating the fitting procedure with nuisance parameters of the respective uncertainties fixed to their post-fit values. The resulting uncertainty is subtracted in quadrature from the total uncertainty. The statistical uncertainty is determined by fixing all nuisance parameters to their post-fit values. Owing to dependencies between nuisance parameters within the fit, quadratic sums of uncertainty contributions can differ from quoted total values. The uncertainty associated to the limited amount of simulated events is quoted apart from the total experimental uncertainty due to their similar contributions

Uncertainty source	$\pm\Delta\mu_{obs}$
Theory	+0.32/−0.30
$t\bar{t}$+hf normalization + parton shower	+0.26/−0.29
Experimental	+0.17/−0.18
b-tagging	+0.13/−0.15
Jet energy corrections	+0.09/−0.09
Amount of simulated events	+0.16/−0.16
Total systematic	+0.43/−0.41
Statistical	+0.26/−0.25
Total	+0.50/−0.48

With a contribution of $+0.43/-0.41$, systematic uncertainties constitute the limiting factor in this analysis. Statistical uncertainties amount to $+0.26/-0.25$, which is consistent with similar analyses at $\sqrt{s} = 13$ TeV [37, 38]. The contribution of experimental uncertainties is $+0.17/-0.18$ with an additional component of $+0.16/-0.16$ accounting for the size of simulated event samples. Uncertainties on b-tagging efficiencies and jet energy corrections denote the leading experimental sources. This can be explained by the fact that only events with a considerable number of jets and b tags pass the selection procedure, which is especially significant in the signal-enriched single-lepton category with at least six jets and at least three b tags. The corresponding effect on the DNN discriminant shape is demonstrated in Figs. 8.4 and 8.5. Therefore, the sensitivity of the analysis on uncertainties related to b-tagging efficiencies and jet energy corrections is increased.

Theory and background modeling uncertainties account for $+0.32/-0.30$ with the dominant contribution of $+0.26/-0.29$ from $t\bar{t}$+hf normalization and parton shower uncertainties. They are evaluated jointly due to interdependencies between the nuisance parameters in the fitting procedure as explained above. To examine the influence of the uncorrelated normalization uncertainty of 50% on each of the four $t\bar{t}$+hf processes (cf. Sect. 8.1.2), the extraction of the best fit value is repeated with changed a-priori uncertainties. For reduced uncertainties of 30% and 20%, this study yields best fit values for μ_{obs} of $0.94\,^{+0.50}_{-0.47}$ and $0.89\,^{+0.47}_{-0.48}$, respectively. While the central values are slightly shifted, uncertainties remain on a comparable level. This finding suggests that the associated nuisances are significantly constrained by the fit, which justifies the initial conservative estimate of 50%.

A comparison of nuisance parameter values and prior distribution widths before and after the fit, and their individual impacts on the best fit value μ_{obs} are shown in Figs. 8.18 and 8.19. Nuisance pulls, i.e., differences between central post- and pre-fit values, divided by their prior distribution widths $\Delta\theta$ are denoted by black markers, with corresponding values shown on the bottom horizontal axis. Accordingly, the ratio between post- and pre-fit distribution widths is represented by black lines. A nuisance is said to be constrained in case this ratio is measured to be below one. Values of nuisance impacts are described by the top horizontal axis. They are computed as the difference between the nominal best fit value μ_{obs} and the value obtained when repeating the fit with the nuisance parameter under scrutiny fixed to its post-fit value plus (red) or minus (blue) its corresponding uncertainty.

Nuisances related to the limited size of simulated event samples are excluded in this representation. As expected, they are only mildly constrained with pulls being normally distributed around zero with a standard deviation of 0.35 ± 0.02. The eight leading impacts are within $\pm[0.02, 0.05]$ and stem from the four right-most bins of the DNN discriminants in the $t\bar{t}H$-enriched categories with single-lepton events containing five and more than six jets. Thus, while the influence of individual nuisances is reasonably small, they contribute to the overall uncertainty due to their quantity (Table 8.4).

Overall, the best fit values of nuisance parameters are found in an interval of $[-1\sigma, +1\sigma]$ around their a-priori expected values. Two outliers with pulls of ±1.5 are related to initial- and final-state radiation uncertainties on the $t\bar{t}+c\bar{c}$ background and presumably result from the estimation of rate-changing effects due to insufficient statistics of systematically varied event samples (cf. Sect. 8.1.2). However, they exhibit impacts of $\lesssim 0.01$ and thus do not affect the measurement of μ_{obs} significantly.

The strongest impact is induced by the $t\bar{t}+b\bar{b}$ normalization uncertainty (Fig. 8.18). Variations of its post-fit value by one standard deviation would cause μ_{obs} to change by ∓0.13. The relative sign of the impact, i.e., a variation by $+1\sigma$ leads to a negative impact, is understandable. The same holds true for the normalization uncertainties on $t\bar{t}+2b$ and $t\bar{t}+c\bar{c}$, although connected to smaller impacts. For $t\bar{t}+b/\bar{b}$, the relative sign is reversed, however, with a reduced absolute impact of ±0.03. Furthermore, with the exception of $t\bar{t}+c\bar{c}$, the uncertainties are constrained up to a factor of 50%, which complies with the examination of fits with reduced a-priori uncertainties as described above.

Uncertainties related to b-tagging efficiencies are significantly constrained owing to the required high b tag multiplicity and the consequently increased sensitivity as explained above (Fig. 8.18). The strongest impact of ±0.07 is related to c-flavor uncertainties, followed by ±0.06 for light-flavor statistics (quadratic), and $+0.03/-0.06$ due to light-flavor contamination in heavy-flavor regions of the b-tagging scale factor measurement.

Nuisance parameters related to initial- and final-state radiation uncertainties, the matching between matrix element and parton showers, and the modeling of the underlying event exhibit slightly larger pulls, but rather small impact values (Fig. 8.18). One exception is the impact of $-0.06/+0.05$ of initial-state radiation uncertainties on $t\bar{t}+$lf background contributions, and to some extent also those for final-state radi-

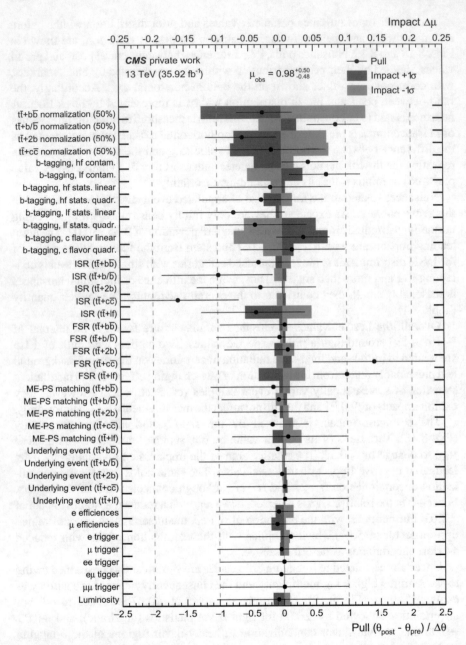

Fig. 8.18 Pulls (black markers) and impacts (red and blue areas) on the best fit value μ of the first 40 nuisance parameters. The values of nuisance pulls are shown on the bottom horizontal axis and are obtained by the difference of post- and pre-fit values θ_{post} and θ_{pre}, respectively, normalized by the pre-fit uncertainty $\Delta\theta$. Black lines correspond to post-fit uncertainties. The top horizontal axis denotes nuisance impacts, which are computed as the difference between the nominal best fit value and the value obtained when performing the fit with the nuisance parameter fixed to its central post-fit value plus (red) or minus (blue) its corresponding uncertainty

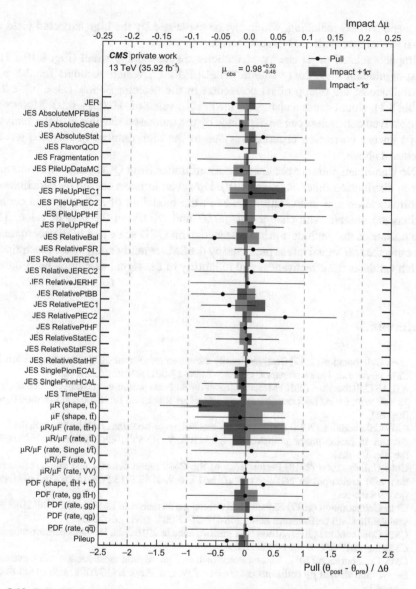

Fig. 8.19 Pulls (black markers) and impacts (red and blue areas) on the best fit value μ of the second 40 nuisance parameters. The values of nuisance pulls are shown on the bottom horizontal axis and are obtained by the difference of post- and pre-fit values θ_{post} and θ_{pre}, respectively, normalized by the pre-fit uncertainty $\Delta\theta$. Black lines correspond to post-fit uncertainties. The top horizontal axis denotes nuisance impacts, which are computed as the difference between the nominal best fit value and the value obtained when performing the fit with the nuisance parameter fixed to its central post-fit value plus (red) or minus (blue) its corresponding uncertainty

ation and ME-PS matching, which can be explained by the high expected yield of $t\bar{t}+$lf events across all analysis categories.

Impacts related to jet energy corrections are reasonably small (Fig. 8.19). The most significant uncertainty source "PileUpPtEC1", which accounts for the p_T-dependence of the pileup offset correction in the detector region $1.3 < |\eta| < 2.5$ (Table 8.1), would cause a change of the best fit value of $+0.06/-0.03$. Moreover, the observed constraints can be explained by the increased sensitivity of the analysis on jet energy correction uncertainties due to the high number of required jets as discussed above.

No significant pulls are obtained for uncertainties from QCD scale variations and parton distribution functions (Fig. 8.19). However, impacts caused by variations of renormalization and factorization scales in the modeling of $t\bar{t}$ background contributions are notable with changes of ∓ 0.07 and ∓ 0.03 on the best fit value. The two nuisances that include a-priori knowledge on QCD scale and PDF uncertainties affecting the $t\bar{t}H$ signal rate as predicted by the SM are neither pulled nor constrained, which validates the effectiveness and reliability of the signal extraction procedure.

References

1. CMS Collaboration (2017) CMS luminosity measurements for the 2016 data taking period. In: CMS physics analysis summary CMS-PAS-LUM-17-001. arXiv:cds:2257069
2. ATLAS Collaboration (2016) Measurement of the inelastic proton-proton cross section at $\sqrt{s} = $ 13 TeV with the ATLAS detector at the LHC. Phys Rev Lett 117:182002. arXiv:1606.02625 [hep-ex]
3. CMS Collaboration (2015) Performance of electron reconstruction and selection with the CMS detector in proton-proton collisions at $\sqrt{s} = 8$ TeV. JINST 10:P06005. arXiv:1502.02701 [physics.ins-det]
4. CMS Collaboration (2018) Performance of the CMS muon detector and muon reconstruction with proton-proton collisions at $\sqrt{s} = 13$ TeV. JINST 13:P06015. arXiv:1804.04528 [physics.ins-det]
5. CMS Collaboration (2017) Electron and photon performance in CMS with the full 2016 data sample. In: CMS performance note CMS-DP-2017-004. arXiv:cds:2255497
6. CMS Collaboration (2017) Muon HLT performance in 2016 data. In: CMS performance note CMS-DP-2017-056. arXiv:cds:2297529
7. CMS Collaboration (2017) Measurement of the $t\bar{t}$ production cross section using events in the $e\mu$ final state in pp collisions at $\sqrt{s} = 13$ TeV. Eur. Phys. J. C 77:172. arXiv:1611.04040 [hep-ex]
8. CMS Collaboration (2019) Measurement of the $t\bar{t}$ production cross section, the top quark mass, and the strong coupling constant using dilepton events in pp collisions at $\sqrt{s} = 13$ TeV. Eur. Phys. J. C 79:368. arXiv:1812.10505 [hep-ex]
9. CMS Collaboration (2019) Measurements of $t\bar{t}$ differential cross sections in proton-proton collisions at $\sqrt{s} = 13$ TeV using events containing two leptons. JHEP 02:149. arXiv:1811.06625 [hep-ex]
10. Cousins RD, Hymes KE, Tucker J (2010) Frequentist evaluation of intervals estimated for a binomial parameter and for the ratio of poisson means. Nucl Instrum Meth A 612:388. arXiv:0905.3831 [physics.data-an]
11. CMS Collaboration (2017) Jet energy scale and resolution in the CMS experiment in pp collisions at 8 TeV. JINST 12:P02014. arXiv:1607.03663 [hep-ex]

12. Sjöstrand T, Mrenna S, Skands P (2006) PYTHIA 6.4 physics and manual. JHEP 05:026. arXiv:hep-ph/0603175
13. Sjöstrand T et al (2015) An introduction to PYTHIA 8.2. Comput Phys Commun 01:024. arXiv:1410.3012 [hep-ph]
14. Bahr M et al (2008) Herwig++ physics and manual. Eur Phys J C 58:639. arXiv:0803.0883 [hep-ph]
15. CMS Collaboration (2013) The performance of the CMS muon detector in proton-proton collisions at $\sqrt{s} = 7$ TeV at the LHC. JINST 8:P11002. arXiv:1306.6905 [physics.ins-det]
16. CMS Collaboration (2018) Identification of heavy-flavour jets with the CMS detector in pp collisions at 13 TeV. JINST 13:P05011. arXiv:1712.07158 [physics.ins-det]
17. Cacciari M, Salam GP (2008) Pileup subtraction using jet areas. Phys Lett B 659:119. arXiv:0707.1378 [hep-ph]
18. Barlow RJ, Beeston C (1993) Fitting using finite Monte Carlo samples. Comput Phys Commun 77:219
19. NNPDF Collaboration (2015) Parton distributions for the LHC Run II. JHEP 04:040. arXiv:1410.8849 [hep-ph]
20. Accardi A et al (2015) The PDF4LHC report on PDFs and LHC data: results from Run I and preparation for Run II. arXiv:1507.00556 [hep-ph]
21. Demartin F et al (2010) The impact of PDF and α_s uncertainties on Higgs Production in gluon fusion at hadron colliders. Phys Rev D 82:014002. arXiv:1004.0962 [hep-ph]
22. Watt G (2011) Parton distribution function dependence of benchmark Standard Model total cross sections at the 7 TeV LHC. JHEP 09:069. arXiv:1106.5788 [hep-ph]
23. Buckley A et al (2015) LHAPDF6: parton density access in the LHC precision era. Eur Phys J C 75:132. arXiv:1412.7420 [hep-ph]
24. Frederix R et al (2012) Four-lepton production at hadron colliders: aMC@NLO predictions with theoretical uncertainties. JHEP 02:099. arXiv:1110.4738 [hep-ph]
25. CMS Collaboration (2016) Event generator tunes obtained from underlying event and multi-parton scattering measurements. Eur Phys J C 76:155. arXiv:1512.00815 [hep-ex]
26. Skands P, Carrazza C, Rojo J (2014) Tuning PYTHIA 8.1: the Monash 2013 tune. Eur Phys J C 74:3024. arXiv:1404.5630 [hep-ph]
27. CMS Collaboration (2016) Investigations of the impact of the parton shower tuning in Pythia 8 in the modelling of $t\bar{t}$ at $\sqrt{s} = 8$ and 13 TeV. In: CMS physics analysis summary CMS-PAS-TOP-16-021. arXiv:cds:2235192
28. Yazgan E (CMS Collaboration) (2017) Top quark modeling and generators in CMS. In: Proceedings of the 10th international workshop on top quark physics, Braga, Portugal. arXiv:1801.05025 [hep-ex]
29. Ježo T, Lindert JM, Moretti N, Pozzorini S (2018) New NLOPS predictions for $t\bar{t} + b$-jet production at the LHC. Eur Phys J C 78:502. arXiv:1802.00426 [hep-ph]
30. CMS Collaboration (2018) Measurements of $t\bar{t}$ cross sections in association with b jets and inclusive jets and their ratio using dilepton final states in pp collisions at $\sqrt{s} = 13$ TeV. Phys Lett B 776:335. arXiv:1705.10141 [hep-ex]
31. Garzelli MV, Kardos A, Trocsanyi Z (2015) Hadroproduction of $t\bar{t}b\bar{b}$ final states at LHC: predictions at NLO accuracy matched with parton shower. JHEP 03:083. arXiv:1408.0266 [hep-ph]
32. Bevilacqua G, Garzelli M, Kardos A (2017) $t\bar{t}b\bar{b}$ hadroproduction with massive bottom quarks with PowHel. arXiv:1709.06915 [hep-ph]
33. Junk T (1999) Confidence level computation for combining searches with small statistics. Nucl Instrum Meth A 434:435. arXiv:hep-ex/9902006
34. Read AL (2002) Presentation of search results: the CL_s technique. J Phys G 28:2693
35. Cowan G, Cranmer K, Gross E, Vitells O (2001) Asymptotic formulae for likelihood-based tests of new physics. Eur Phys J C 71:1554. arXiv:1007.1727 [physics.data-an]
36. LHC Higgs Cross Section Working Group (2017) Handbook of LHC Higgs cross sections: 4. deciphering the nature of the Higgs sector. In: CERN yellow reports: monographs. arXiv:1610.07922 [hep-ph], cds:2227475

37. ATLAS Collaboration (2018) Search for the standard model Higgs boson produced in association with top quarks and decaying into a $b\bar{b}$ pair in pp collisions at $\sqrt{s} = 13$ TeV with the ATLAS detector. Phys Rev D 97:072016. arXiv:1712.08895 [hep-ex]
38. CMS Collaboration (2019) Search for $t\bar{t}H$ production in the $H \rightarrow b\bar{b}$ decay channel with leptonic $t\bar{t}$ decays in proton-proton collisions at $\sqrt{s} = 13$ TeV. JHEP 03:026. arXiv:1804.03682 [hep-ex]
39. CMS Collaboration (2015) Search for a standard model Higgs boson produced in association with a top-quark pair and decaying to bottom quarks using a matrix element method. Eur Phys J C 75:251. arXiv:1502.02485 [hep-ex]
40. ATLAS Collaboration (2015) Search for the standard model Higgs boson produced in association with top quarks and decaying into $b\bar{b}$ in pp collisions at $\sqrt{s} = 8$ TeV with the ATLAS detector. Eur Phys J C 75:349. arXiv:1503.05066 [hep-ex]

Chapter 9
Conclusion

After the discovery of the Higgs boson during the first run of the LHC, precise measurements of its properties and couplings to other particles are conducted by the ATLAS and CMS experiments. Due to its high mass, the Standard Model predicts that the top quark, among all fermions, exhibits the strongest coupling to the Higgs boson with a Yukawa coupling constant close to unity. Therefore, the accurate measurement of Higgs-top coupling is essential for probing predictions of the SM and to constrain theories that reach beyond it.

In this thesis, a search for the associated production of a Higgs boson with a top quark pair at a center-of-mass energy of $\sqrt{s} = 13$ TeV was presented. The data from proton-proton collisions of the second run of the LHC was recorded by the CMS detector in 2016 and corresponds to an integrated luminosity of $35.9\,\text{fb}^{-1}$. The analysis focused on events where the Higgs boson decays into a pair of bottom quarks whereas the decay of the $t\bar{t}$ system involves one (single-lepton) or two (dilepton) electrons or muons with opposite charge. In addition to the lepton requirements, events were selected to have missing transverse energy and at least four energetic jets of which at least three must have a b tag. The parameters of the event selection were chosen to optimize the yield of expected signal events while suppressing background contributions. Furthermore, the criteria were synchronized with CMS analyses of further $t\bar{t}H$ final states in order to avoid double-counting of events during the combination of results.

The analysis was dominated by systematic uncertainties in the normalization of $t\bar{t}$ background contributions with additional heavy-flavor jets, especially from $t\bar{t}+b\bar{b}$, $t\bar{t}+b/\bar{b}$, $t\bar{t}+2b$, and $t\bar{t}+c\bar{c}$ production. To reduce their impact on the analysis, a novel event categorization scheme based on deep neural networks was introduced. In contrast to binary classification, the networks performed a multi-class classification, which attributed events with a probability to originate from either the signal process or a particular background process. The information of the highest probability was

© The Author(s), under exclusive license to Springer Nature Switzerland AG 2021
M. Rieger, *Search for t̄tH Production in the H → bb̄ Decay Channel*, Springer Theses,
https://doi.org/10.1007/978-3-030-65380-4_9

exploited to create mutually exclusive categories that were enriched with events of one of the considered physics processes. As the $t\bar{t}$+heavy-flavor normalization uncertainties were treated as uncorrelated in the signal extraction procedure, the simultaneous isolation of backgrounds using this approach led to more accurate background constraints and, in turn, to a gain in signal sensitivity.

Besides thorough monitoring of the neural network output and training process, a new technique for validating input variables was established. Since deep neural networks are able to exploit deep correlations, commonly used one-dimensional comparisons between distributions of simulated and recorded event variables are insufficient to assess their agreement. The introduced method also covers the validation of higher-order correlations. Two-dimensional distributions of all pairs of various simple and composite variables were fitted to data, whereby the fit qualities were assessed through saturated goodness-of-fit tests. Only variables that yielded high p-values in combination with most of the other variables were considered as input variables.

The scale and complexity of the analysis were primarily driven by the amount of collision data to be processed, the data flow entailed by the utilized machine learning approach, and the treatment of a large number of systematic uncertainties. Technical considerations were necessary to organize the analysis in a way that ensured an efficient execution across distributed infrastructure. Therefore, a novel workflow management system was developed, which establishes a generic analysis design pattern with support for interchangeable remote resources.

Results were obtained in a simultaneous measurement over 24 orthogonal categories defined by the lepton channel, the number of jets, and the most probable process as attributed by the deep neural networks. In each category, the distribution of the network output unit that corresponds to the assigned process category was fitted to data. Upper limits on the $t\bar{t}H$ signal strength modifier $\mu = \sigma/\sigma_{SM}$ were computed using the asymptotic CL_S method at 95% confidence level. The expected limit expresses the compatibility of the SM prediction with the background-only hypothesis, whereas the observed limit states the signal strength threshold above which $t\bar{t}H$ production is excluded with 95% confidence. The best fit value on μ was extracted in a profile likelihood fit. Systematic uncertainties were incorporated as nuisance parameters in the fitting procedure.

The SM predicts a $t\bar{t}H$ production cross section of $\sigma_{SM} = 507^{+35}_{-50}$ fb with NLO QCD accuracy and NLO electroweak corrections [1]. For a Higgs boson mass of $m_H = 125$ GeV, this analysis measured an observed upper limit on the $t\bar{t}H$ signal strength modifier of

$$\mu_{obs} < 1.85, \tag{9.1}$$

which excludes greater values with 95% confidence given the measured data. The background-only hypothesis is disfavored due to an expected limit of $\mu_{exp} < 0.99^{+0.42}_{-0.29}$ in case of a signal strength as predicted by the SM. The observed best fit value μ_{obs} was measured as

$$\mu_{obs} = 0.98^{+0.50}_{-0.48} = 0.98^{+0.26}_{-0.25} \text{ (stat.)} ^{+0.43}_{-0.41} \text{ (syst.)} \tag{9.2}$$

with uncertainties stated in units of σ_{SM}, leading to the $t\bar{t}H$ production cross section of

$$\sigma_{t\bar{t}H} = 497 \, ^{+252}_{-243} \, \text{fb} = 497 \, ^{+130}_{-128} \, (\text{stat.}) \, ^{+216}_{-206} \, (\text{syst.}) \, \text{fb}. \quad (9.3)$$

A value of $\mu_{exp} = 1.00^{+0.55}_{-0.49}$ was expected from simulation. The corresponding observed (expected) significance amounts to 2.04 (1.92) standard deviations above the expected background. The measured value is in agreement with the SM as well as with a search performed by the ATLAS experiment using a similar amount of proton-proton collision data at a center-of-mass energy of $\sqrt{s} = 13\,\text{TeV}$ [2]. With an uncertainty of $+51/-49\%$ relative to $\sigma_{t\bar{t}H}$, however, the presented analysis is considerably more accurate.

The presented analysis in the single-lepton channel contributed to the first observations of $t\bar{t}H$ production [3] and Higgs bosons decaying into bottom quarks [4], performed with the CMS experiment. In addition, it was published in combination with an alternative strategy in the dilepton channel in Ref. [5]. With a best fit value of $\mu = 0.72 \pm 0.45$, the published result is comparable to that stated above regarding both its central value and accuracy.

The total uncertainty on μ_{obs} of $+0.50/-0.48$ predominantly originates from systematic effects. With a contribution of $+0.26/-0.29$, the uncertainty on $t\bar{t}$+heavy-flavor background normalizations constitutes the most significant source of systematic uncertainty. In comparison, experimental uncertainties, which are mainly driven by uncertainties related to b-tagging, and those resulting from the limited size of simulated event samples, amount to uncertainties of $+0.13/-0.15$ and $+0.16/-0.16$, respectively.

Further analyses may benefit from the strategy to employ deep neural networks for separating signal and background contributions via multi-class classification as developed in this thesis. Physics-motivated network architectures, refined theoretical models describing $t\bar{t}$+heavy-flavor production, and the utilization of new jet-flavor-tagging algorithms are expected to have a significant impact on following iterations. The precise measurement of $t\bar{t}H$ production will remain an essential goal of the LHC physics program in order to understand the nature of electroweak symmetry breaking. It will enable future analyses involving, for example, Higgs boson self-coupling and single tH production, which are already on the horizon.

References

1. LHC Higgs Cross Section Working Group (2017) Handbook of LHC Higgs cross sections: 4. deciphering the nature of the Higgs sector. CERN Yellow reports: monographs. http://arxiv.org/abs/1610.07922, http://cds.cern.ch/record/2227475
2. ATLAS Collaboration (2018) Search for the standard model Higgs boson produced in association with top quarks and decaying into a $b\bar{b}$ pair in pp collisions at $\sqrt{s} = 13$ TeV with the ATLAS detector. Phys Rev D 97:072016. arXiv:1712.08895 [hep-ex]
3. CMS Collaboration (2018) Observation of $t\bar{t}H$ production. Phys Rev Lett 120(23):231801. arXiv:1804.02610 [hep-ex]

4. CMS Collaboration (2018) Observation of Higgs boson decay to bottom quarks. Phys Rev Lett 121:121801. arXiv:1808.08242 [hep-ex]
5. CMS Collaboration (2019) Search for $t\bar{t}H$ production in the $H \rightarrow b\bar{b}$ decay channel with leptonic $t\bar{t}$ decays in proton-proton collisions at $\sqrt{s} = 13$ TeV. JHEP03:026. arXiv:1804.03682 [hep-ex]

Appendix A
DNN Input Variables

The input variable distributions presented in the following are listed in Tables 7.1 and 7.2. They are evaluated for single-lepton events with at least six jets. The agreement between data and simulation is comparable for single-lepton events with four and five jets, and for dilepton events with at least four jets (Figs. A.1, A.2, A.3 and A.4).

© The Editor(s) (if applicable) and The Author(s), under exclusive license to Springer   207
Nature Switzerland AG 2021
M. Rieger, *Search for t t̄ H Production in the H* ⟶ *b b̄ Decay Channel*, Springer Theses,
https://doi.org/10.1007/978-3-030-65380-4

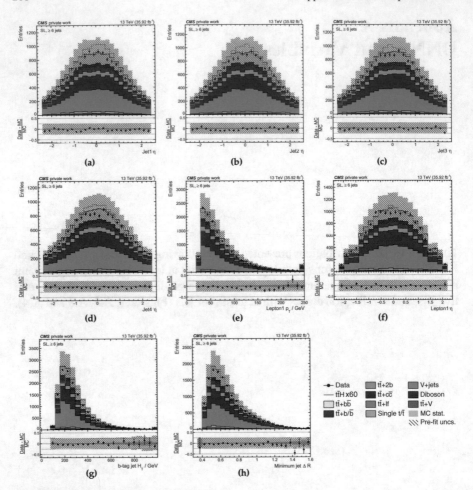

Fig. A.1 Distributions of variables that are considered as input to deep neural networks for single-lepton events with at least six jets. The $t\bar{t}H$ signal is shown superimposed and scaled by a factor of 60 for visibility. The filled gray areas in the ratio plots denote statistical uncertainties due to the limited number of simulated events, whereas hatched areas describe combined systematic uncertainties on expected background contributions, prior to the fit to data

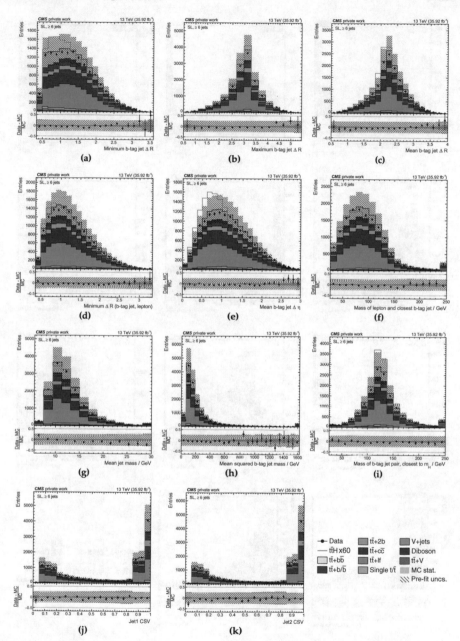

Fig. A.2 Distributions of variables that are considered as input to deep neural networks for single-lepton events with at least six jets. The $t\bar{t}H$ signal is shown superimposed and scaled by a factor of 60 for visibility. The filled gray areas in the ratio plots denote statistical uncertainties due to the limited number of simulated events, whereas hatched areas describe combined systematic uncertainties on expected background contributions, prior to the fit to data

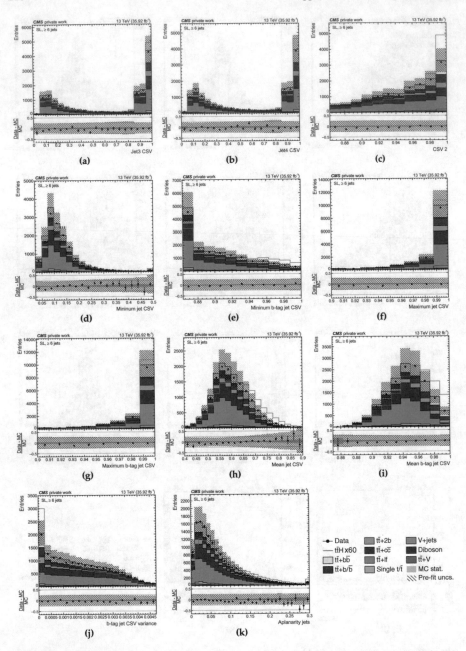

Fig. A.3 Distributions of variables that are considered as input to deep neural networks for single-lepton events with at least six jets. The $t\bar{t}H$ signal is shown superimposed and scaled by a factor of 60 for visibility. The filled gray areas in the ratio plots denote statistical uncertainties due to the limited number of simulated events, whereas hatched areas describe combined systematic uncertainties on expected background contributions, prior to the fit to data

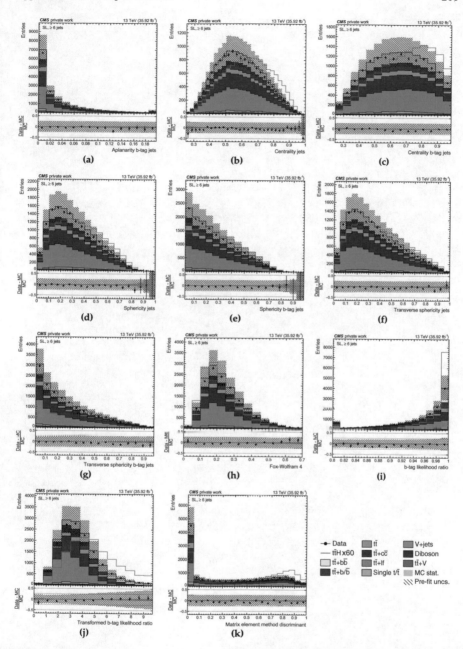

Fig. A.4 Distributions of variables that are considered as input to deep neural networks for single-lepton events with at least six jets. The $t\bar{t}H$ signal is shown superimposed and scaled by a factor of 60 for visibility. The filled gray areas in the ratio plots denote statistical uncertainties due to the limited number of simulated events, whereas hatched areas describe combined systematic uncertainties on expected background contributions, prior to the fit to data

Appendix B
DNN Output Distributions

See Figs. B.1, B.2, B.3 and B.4.

M. Rieger, *Search for tt̄H Production in the H* ⟶ *bb̄ Decay Channel*, Springer Theses,
https://doi.org/10.1007/978-3-030-65380-4

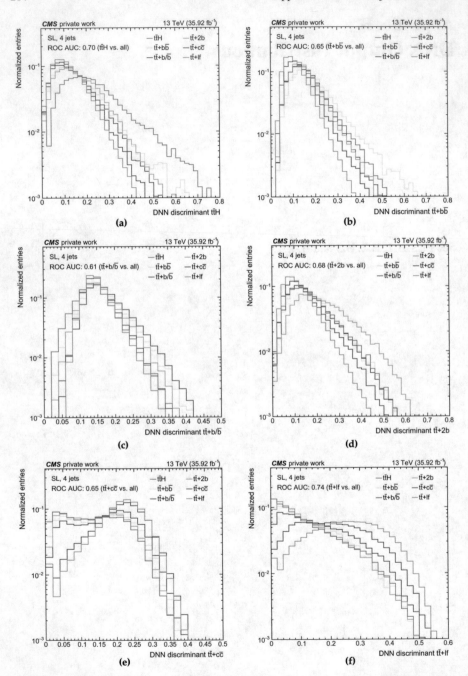

Fig. B.1 Distributions of the six output units of the deep neural network for single-lepton events with four jets, namely $t\bar{t}H$ (**a**), $t\bar{t}+b\bar{b}$ (**b**), $t\bar{t}+b/\bar{b}$ (**c**), $t\bar{t}+2b$ (**d**), $t\bar{t}+c\bar{c}$ (**e**), and $t\bar{t}+$lf (**f**). The output of each unit is shown separately for $t\bar{t}H$ and $t\bar{t}$ processes. The value range is adjusted individually per unit for visualizing shape differences

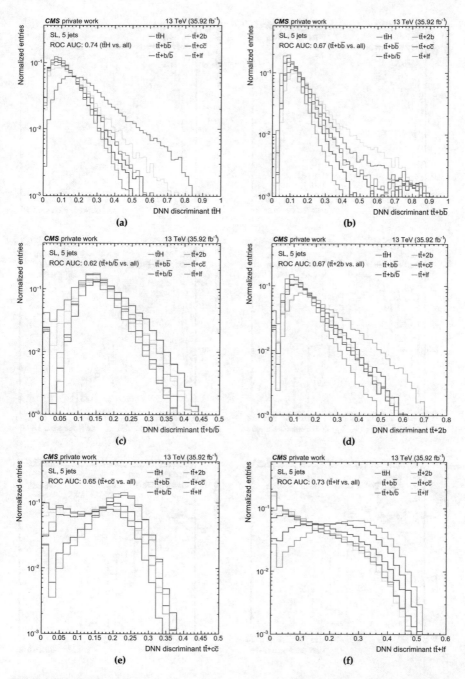

Fig. B.2 Distributions of the six output units of the deep neural network for single-lepton events with five jets, namely $t\bar{t}H$ (**a**), $t\bar{t}+b\bar{b}$ (**b**), $t\bar{t}+b/\bar{b}$ (**c**), $t\bar{t}+2b$ (**d**), $t\bar{t}+c\bar{c}$ (**e**), and $t\bar{t}+lf$ (**f**). The output of each unit is shown separately for $t\bar{t}H$ and $t\bar{t}$ processes. The value range is adjusted individually per unit for visualizing shape differences

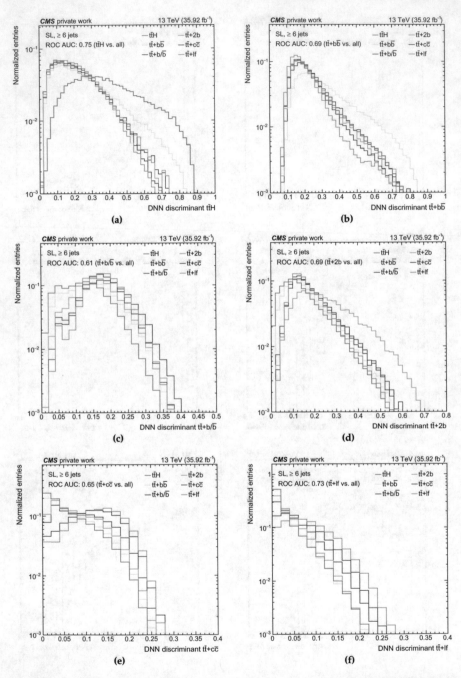

Fig. B.3 Distributions of the six output units of the deep neural network for single-lepton events with at least six jets, namely $t\bar{t}H$ (**a**), $t\bar{t}+b\bar{b}$ (**b**), $t\bar{t}+b/\bar{b}$ (**c**), $t\bar{t}+2b$ (**d**), $t\bar{t}+c\bar{c}$ (**e**), and $t\bar{t}+$lf (**f**). The output of each unit is shown separately for $t\bar{t}H$ and $t\bar{t}$ processes. The value range is adjusted individually per unit for visualizing shape differences

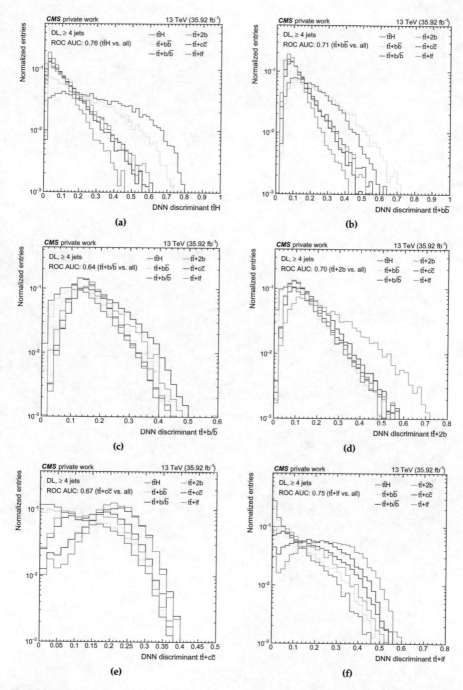

Fig. B.4 Distributions of the six output units of the deep neural network for dilepton events with at least four jets, namely $t\bar{t}H$ (**a**), $t\bar{t}+b\bar{b}$ (**b**), $t\bar{t}+b/\bar{b}$ (**c**), $t\bar{t}+2b$ (**d**), $t\bar{t}+c\bar{c}$ (**e**), and $t\bar{t}+$lf (**f**). The output of each unit is shown separately for $t\bar{t}H$ and $t\bar{t}$ processes. The value range is adjusted individually per unit for visualizing shape differences

Printed in the United States
by Baker & Taylor Publisher Services